Textile Science and Clothing Technology

Series Editor

Subramanian Senthilkannan Muthu, SgT Group & API, Hong Kong, Kowloon, Hong Kong

This series aims to broadly cover all the aspects related to textiles science and technology and clothing science and technology. Below are the areas fall under the aims and scope of this series, but not limited to: Production and properties of various natural and synthetic fibres; Production and properties of different yarns, fabrics and apparels; Manufacturing aspects of textiles and clothing; Modelling and Simulation aspects related to textiles and clothing; Production and properties of Nonwovens; Evaluation/testing of various properties of textiles and clothing products; Supply chain management of textiles and clothing; Aspects related to Clothing Science such as comfort; Functional aspects and evaluation of textiles; Textile biomaterials and bioengineering; Nano, micro, smart, sport and intelligent textiles; Various aspects of industrial and technical applications of textiles and clothing; Apparel manufacturing and engineering; New developments and applications pertaining to textiles and clothing materials and their manufacturing methods; Textile design aspects; Sustainable fashion and textiles; Green Textiles and Eco-Fashion; Sustainability aspects of textiles and clothing; Environmental assessments of textiles and clothing supply chain; Green Composites; Sustainable Luxury and Sustainable Consumption; Waste Management in Textiles; Sustainability Standards and Green labels; Social and Economic Sustainability of Textiles and Clothing.

More information about this series at https://link.springer.com/bookseries/13111

Xinfeng Yan · Lihong Chen · Hafeezullah Memon
Editors

Textile and Fashion Education Internationalization

A Promising Discipline from South Asia

Springer

Editors
Xinfeng Yan
International Cultural Exchange School
Donghua University
Shanghai, China

Lihong Chen
Shanghai International Fashion Science and Innovation Center
Donghua University
Shanghai, China

Hafeezullah Memon
College of Textile Science and Engineering
Zhejiang Sci-Tech University
Hangzhou, Zhejiang, China

We acknowledge the National Social Science Foundation of China (BGA200057), Research Fund for International Scientists (RFIS-52150410416), National Natural Science Foundation of China, as well as Research Startup Grant of ZSTU (20202294-Y).

ISSN 2197-9863 ISSN 2197-9871 (electronic)
Textile Science and Clothing Technology
ISBN 978-981-16-8853-9 ISBN 978-981-16-8854-6 (eBook)
https://doi.org/10.1007/978-981-16-8854-6

This Springer imprint is published by the registered company Springer Nature Singapore Pte Ltd.
The registered company address is: 152 Beach Road, #21-01/04 Gateway East, Singapore 189721, Singapore

Foreword

South Asia, with its population of over 1.8 billion, not only provides a massive internal market for textile goods of all kinds but also makes a considerable contribution to textiles and fashion globally. From the 1980s and 1990s through the early years of the twenty-first century until the global financial crisis in 2008, the Chinese textile and clothing industries attracted much more attention than those in South Asia, even though there was significant participation by Bangladesh, India, Nepal, Pakistan, and Sri Lanka in the globalization of garment production. China can still act as a beacon for South Asia, not only through the experiences gained from the development of its textile and clothing industries to supply the growing global market for fashion goods but also for its demonstration of how to switch attention to the domestic market at the time of a financial crisis; in that way, it ensured the industry's continued contribution to high levels of employment and the country's economy. China's technical and scientific support for its industry was predominantly provided by the universities, polytechnics, and colleges from within China itself and Hong Kong SAR, and over that rapid-development period many of these institutions have developed worldwide recognition.

For South Asia, for its textile and clothing industries to perform domestically and globally to best effect, its universities, colleges, and research institutes have a key role to play. This book is written by academics from within those higher education institutions, and they provide others such as myself with an insight into the challenges they face and the contributions they intend to make in the future to their industry and their country's economy. It is fair to say that they will face challenges from other parts of higher education, particularly from those devoted to studying traditional subjects such as mathematics, physics, chemistry, the arts, language, and culture. In the eyes of some, it is the study of these subjects alone that deserves the highest academic recognition. However, typically, that is simply because they themselves have insufficient vocational education experience (what is described here as "application-oriented talent education"). In practice, what was widely found in Western countries from the 1880s onward, was that to move applied science forward with the required rapidity in support of the industrial revolution, improved educational provision that was intensely focused on the industry's technological development needs was essential. The same applied to design education. It is important for those who pursue the study of, and/or teach traditional academic subjects, to be helped to realize that just because those involved in vocational research and education focus their attention on a particular aspect of applied science or design, it does not mean that their intelligence is then lost or damaged in some way. Examples given here show that it is also important for the institutions involved in vocational higher education to win the support of the industry trade bodies for the research and the education they provide and for the contribution made to the industry by their graduates. Professional bodies can also assist in gaining proper recognition.

Each chapter in this book enlightens the reader about the approach taken to textile and clothing education in different parts of South Asia. In some cases, research and vocational higher education are well-developed, and the academic staff involved play significant roles not just in their own countries but also on the world stage. In others, the development of the textile industry and related vocational education are on the cusp between gaining understanding and recording craft traditions and designs and moving at least some of that heritage into the beginnings of an industry (the Maldives), while others are focused on supporting the exercise of traditional craft skills through practice-based training (Bhutan). In all cases, however, there is recognition of the continued significance in the domestic market for traditional clothing items alongside those that follow international trends. The challenge is in how to take advantage of both. A good example is that of Nepal, where there was strong industry support for educational improvement resulting in the establishment of its textile higher education. Its graduates built on Nepal's early participation in the global fashion market with the development of a strong fashion-design link with Bollywood and have been able to set this alongside the development of what is known there as neo-modern fashion.

The important contribution of textiles and textile higher education to the economies of India, Pakistan, Bangladesh, and Sri Lanka and the way each of them benefits from government recognition are demonstrated. There are well-established arrangements for textile higher education in India, and its various subdivisions are explained, as are the control arrangements established to maintain quality of provision. There is a useful listing of institutes/institutions providing higher education to all levels up to research degrees and of research associations. Overall, the indication is that there is recognition in the government of the contribution made by its textile academics and strong government support for textile-related higher education. The same applies in Pakistan, and a useful listing and some of the history of their development is provided for both public and private textile educational institutes. The focus on learning outcomes in taught programs in its textile universities is explained, as are the benefits to be gained from that approach, alongside the application of student-support mechanisms and industrial liaison. In Bangladesh, their ready-made garment industry is the largest single source of foreign currency and second only to China, so the government shows a strong interest in maintaining that position. One of the primary goals in the nation's educational policy has been to establish a suitable environment for textile studies and related disciplines with a strong focus on textile engineering in universities.

Again a useful list of public and private institutions providing courses on disciplines related to textile engineering is given; there is provision at all levels, including research, but teacher shortages are said to be a major inhibiting factor. There is also still considered to be a skills gap between those held by emerging graduates compared with industry requirements. Efforts are having to be made to improve the availability of suitably sized machines to gain the missing expertise, so study-abroad and international inter-university collaboration opportunities are also examined. For Sri Lanka, three stakeholder groups are required to work together in human resource development for the nation's recently restructured apparel and textile industry. It is explained that these are intended to form the necessary bridge between industry, universities, and government. There is insufficient capability in weaving to provide a domestic supply of woven fabrics for the garment manufacturing industry, so some correction needs to be applied there, and there is also said to be a need to enable their fashion designers and design students to become more familiar with cutting-edge technology. The experiences of the more-developed countries in relation to their textiles industries and textile-related higher education, India, Pakistan, Bangladesh, and Sri Lanka vary considerably, but together, they can offer ideas and solutions for others to consider, or at least to consider in part, not least for Afghanistan, where the potential is considerable, but the full vocational educational infrastructure needs to be established.

Prof. Richard Murray, CompTI, B.Sc.,
Ph.D., CText FTI
Emeritus Professor
Manchester Metropolitan University
Manchester, UK

The original version of the book was revised: For detailed information please see Correction. The correction to the book is available at https://doi.org/10.1007/978-981-16-8854-6_10

Preface

South Asian countries are one of the central regions of textile and garment production and processing. Their population advantages and cost advantages provide dividends for the development of textile manufacturing. Therefore, there is a strong demand for talents training in the textile and garment industry. In the supply of Applied Talents in the textile industry, we need to learn and exchange with experienced countries (China). Therefore, it is vital to strengthen talents in the textile and garment industry and promote international exchange and cooperation in higher education. Specifically, the collaboration between universities with textile industry characteristics can be carried out in the development trend of textile education, internationalization of textile education, talent training mode of textile education, and teaching reform of textile education.

In talent training in the textile industry, more and more countries pay attention to application-oriented education. South Asian countries regard the development of applied talents education as an essential strategy to improve human resources. Generally speaking, there are some common problems in the development of application-oriented education in South Asian countries, which are mainly reflected in the following four aspects:

1. The low status of application-oriented talent education;
2. The insufficient investment in education funds;
3. The need to improve the quality of education;
4. The urgent need to improve education management.

First of all, the education status of applied talents is low. In South Asian countries, the popularization rate of compulsory education is low. The labor force only has the education level of primary school or below, making it difficult for many workers to enter the corresponding educational institutions to receive applied education and training. At the same time, influenced by the traditional ideas, many people are unwilling to engage in skilled jobs and are more inclined to work in government departments or become white-collar workers. Secondly, the investment in education is insufficient. The economic development level of South Asian countries is relatively backward, and the per capita income is low, which leads to a severe shortage of overall investment in education. Third, the quality of education needs to be improved. Low quality is a common problem in higher education in South Asian countries. One of the root causes of the problem lies in the low level of teaching staff and the lack of practical skills. Fourth, education management needs to be improved.

On the one hand, as far as applied education management institutions are concerned, the scientificness, rationality, and fairness of educational development policies need to be strengthened. The pertinence of relevant policies also needs to be maintained. On the other hand, the contact and communication mechanism between education management institutions and industry and human resource management departments must be improved.

Through the analysis of the current situation and development trend of textile and garment education in South Asian countries, this book explores the paths and methods of international exchange and cooperation of higher education with industry characteristics to provide the analytical basis and theoretical support for the practice of international collaboration between colleges and universities. We have welcomed to describe the diverse aspects to emphasize different aspects of textile education of South Asia to retain the interest and attention of a large number of readers in this book. For example, Chapter "Textile Academics in India—An Overview" lists and discusses Indian fabrics, academic programs and institutes, research associations, and schemes under the textile ministry. Chapter "Brief Analysis on the Past, Present, and Future of Textile Education in Bangladesh" emphasizes the history of textile education, employment, and the existing skill gap in Bangladesh's textile and apparel industry. More importantly, it elaborates textile education level, present and future projects, and scopes for textile students on higher education or research opportunities. Chapter "Textile Education in Pakistan" focuses on outcome-based education and assessment methods in the textile academia after discussing a brief history of textile education in Pakistan. In addition, admission and resources for students and some local scholarships to pursue textile higher education are also presented. Chapter "Prospects of Textile Education in Afghanistan" discusses the background of the textile industry, related commodities' production, and prospects of the textile industry and academia in Afghanistan. Chapter "Namuna College of Fashion Technology: Pioneering in Fashion and Textile Education in Nepal" presents the history, opportunities and challenges, curricular structure, and alumni of NCFT as pioneering in fashion and textile education in Nepal. Textile education in the early administrative and pre-independence era, including regenerating traditional industries in the national context liberalization of the Sri Lankan economy and textile education toward the twenty-first century, is presented in Chapter "Evolution of Sri Lankan Textile Education from Ancient Times to the 21st Century". The textile heritage of Maldives, particularly some specific types of weaves, is nicely presented in the Chapter "Textiles as Heritage in the Maldives". Textile education in Bhutan is transferred from generation to generation. The changes in the textile education generation are covered adequately by surveying and interviewing three generations in the Chapter "Weaving Through Generations: A Study on the Transmission of Bhutanese Weaving Knowledge and Skills Over Three Generation".

Thank you for our experts' knowledge, deep analysis, and wise judgment. As the editors of this book, our work is to integrate these research results and present these wisdom crystals to the readers in the professional field to maximize the sharing and utilization of knowledge value.

Shanghai, China — Xinfeng Yan
Shanghai, China — Lihong Chen
Hangzhou, China — Hafeezullah Memon

Contents

Introduction

Xinfeng Yan, Lihong Chen, and Hafeezullah Memon

Abstract This book chapter describes the general overview of higher education in South Asia and its internationalization. Moreover, this chapter focuses on the specific problems of textile and clothing education internationalization in South Asian countries for comparative analysis and in-depth discussion, taking the international cooperation and development of talent education in textile and clothing as an example. It can provide new ideas for the international exchange and cooperation of education in the field of textile and clothing in the region through the analysis of the current situation, experience, and trend of the development of higher education and international cooperation in the field of textile and clothing in various countries in South Asia. The vital foundation and premise of the internationalization of higher education are the coordinated development of industry, the dynamic allocation of resources, and the cross-demand of talents.

Keywords South Asia · Internationalization · Textile · Economy · Higher Education

1 International Cooperation and Development of Education is the Result of Economic Globalization

Under the background of globalization, international cooperation and development of education are more and more frequent. International exchange and cooperation in

X. Yan
Donghua University, Shanghai 200051, China

L. Chen
Shanghai International Fashion Science and Innovation Center, Donghua University, Shanghai 200051, China

H. Memon (✉)
College of Textile Science and Engineering, International Institute of Silk, Zhejiang Sci-Tech University, Hangzhou 310018, China
e-mail: hm@zstu.edu.cn

X. Yan et al. (eds.), *Textile and Fashion Education Internationalization*, Textile Science and Clothing Technology, https://doi.org/10.1007/978-981-16-8854-6_1

education is a crucial way to enhance cultural exchange between countries. Cooperation and development in the field of economy and trade need to be based on the understanding of cultural exchanges. Therefore, educational cooperation is of great significance to the international development of the regional economy, and it is an essential basis for enhancing understanding, identity, and win-win cooperation [1]. The internationalization of higher education integrates international cross-cultural and global school running ideas, governance mechanisms, and management modes into universities' goal, function, and innovative culture construction. It emphasizes the efforts of the state and higher education institutions to improve the quality of personnel training and scientific research, improve the organization and operation of universities, promote the development of higher education, and continuous introduction of international resources for self-transformation [2]. The internationalization of higher education has become the core force to promote the reform and development of higher education [3, 4].

Economic globalization is the main driving force for the internationalization of higher education. South Asia includes India, Pakistan, Bangladesh, Nepal, Sri Lanka, Bhutan, Afghanistan, and the Maldives, with vast territory, large population, and rich resources. On the whole, South Asia is one of the fastest-growing regions in the world [5, 6]. In 2016, the economic growth rate of South Asia was 6.7%, much higher than that of most other regions, and the regional economic growth rate ranked first in the world. South Asia has a large population, with more than 1.8 billion people in 2018. It is also one of the four most densely populated regions in the world. The vast population base provides a realistic foundation for the internationalization of education. At the same time, the overall level of education in South Asia is low. Therefore, it is also a meaningful way to organize educational exchanges and cooperation and introduce external resources. Regional exchanges and cooperation are equally crucial to the international development of education. The South Asian Association for Regional Cooperation (SAARC) is a regional mutual aid and cooperation organization established by South Asian countries to meet international development needs [7]. Its purpose is to strengthen exchanges between countries and promote win-win cooperation in society, culture, technology, and science.

The internationalization of education will also have an impact on international cooperation in the economic field. International education has a significant positive impact on the economic performance of studying abroad countries [8, 9]. Therefore, in Pakistan as well as India, there are many scholarships to attract international students. Although both countries are developing countries, they offer numerous scholarships to international students, opening the door of the foreign economy.

2 The Current Situation and Characteristics of Higher Education in South Asian Countries

The internationalization of higher education includes the value orientation of higher education, the idea of running a school, the way of realization, the composition of students, the flow of teachers, curriculum, etc. The characteristics and advantages of higher education in different countries are different, so the specific internationalization practice will also reflect different purposes and characteristics. For example, South Asian countries are different in politics, economy, culture, and policy orientation. Consequently, their higher education presents different characteristics.

India is the highland of higher education in South Asia, originating from the British colonial rule, with distinctive characteristics of the British education model. Indian higher education has a long history and a large scale. India has the highest number of universities globally; according to last year's statistics, there are 4381 comprehensive universities [8] and 39,931 colleges [9]. The nature of colleges and universities is mainly private, and the government funds only about 30% of them. The higher share of private institutes suggests the interest of local people in higher education. Regarding educational forms, the proportion of Open University and distance education is relatively high, about 25%, and the demand is still growing.

Nevertheless, India's existing education resources can only meet about 10% of the population of the right age, far below the average level of developing countries. Moreover, the general framework of the general agreement on trade and services (GATS) sets rules for the operation of transnational higher education. After India signed the GATS, more than 100 foreign universities have been running schools in India.

Pakistan's higher education started late and adopted the British education model after independence. When it was founded in 1947, there was only one Punjab University in Lahore. Overall, the development of higher education in Pakistan is relatively slower than in India, the number of universities is relatively small, and public universities are the main ones. The traditional dual track higher education management system is adopted. In 2016, there were 91 (56%) public sector and 72 (44%) private sector, total 163 universities, 1418 degree-awarding colleges, including 1259 (89%) public and 159 (11%) private; and 3746 technical and vocational institutions, including 1123 (30%) are public institutions and 2623 (70%) are private institutions in Pakistan [10]. The gross enrollment rate is still less than 15%, which is in the elite stage of higher education. Educational resources are insufficient, and even the number of teachers has a downward trend in a certain period. The number of students studying abroad is increasing year by year, indicating a phenomenon of talent flow and even loss in Pakistan's Higher Education [11].

There are 34 public universities and 56 private universities in higher education institutions in Bangladesh. Public universities mainly follow the British system's three-year bachelor's degree mode, while private universities mainly adopt the four-year undergraduate course mode of the United States. Public universities are free of charge and funded by the government. However, due to the lack of economic pressure on public universities, there is a lack of close connection between talent cultivation and the employment market, and the mismatch between graduates' skills and market demand is serious. On the other hand, private higher education institutions charge higher fees, mainly for the children of wealthy families, and their curriculum is relatively flexible and can meet the market demand, but the guarantee of teaching quality is a big problem.

The development of higher education in Nepal can be traced back to the establishment of Tri-Chandra College in 1918, the first higher education institution in Nepal. There are 15 universities and 1407 colleges in Nepal. Although the government encourages private schools to participate in higher education to solve the shortage of educational resources, most private institutions do not have enough infrastructure support. Most occupy rental housing, hire part-time teachers to attend classes, and lack standardized academic activities. Due to the shortage of domestic higher education resources, studying abroad is becoming a significant choice for Nepalese students, and the number of students studying abroad is increasing year by year.

Bhutan's higher education began in 1983 when Sherubtse Junior College was upgraded from a public tertiary school in Kanglung, Trashigang district. At present, only the Royal University of Bhutan is a comprehensive university. Most Bhutanese students have overseas training experience, including India, Singapore, Japan, Australia, New Zealand, the United Kingdom, Germany, and the United States. Most Bhutanese students return home after finishing their studies.

Influenced by the domestic political situation, Sri Lanka's higher education started late and developed slowly. The British education system profoundly influences the education system. After the end of the twenty-six-year civil war in 2009, Sri Lanka formally put forward the "Mahinda vision" of national development to build Sri Lanka into an Asian knowledge, aviation, investment, business, and energy center. The "strategic management plan" in higher education is a part of the "Mahinda Chintana vision", which mainly includes expanding the number of educations, improving the quality of courses, enhancing international exchanges, and other aspects. Sri Lanka implements the free education policy, but the investment in education is insufficient, and the proportion of education expenditure in GDP is among the lowest in the world. As a result, higher education enrollment is low, and the professional structure is unreasonable, restricting economic development. The higher entrance threshold of public universities leads to strong demand for private universities and overseas students. Most students prefer to study in Universities in the U.K., Australia, Malaysia, the United States, Singapore, and other countries.

The Maldives has a small territory, a small population, and a high literacy rate of about 99%. Primary and secondary education is free and compulsory, but the development of higher education is relatively late. Before 1998, there were no fundamental higher education institutions. Higher education institutions are mainly composed of the Maldives Institute of higher education, Islamic Research Institute, continuing education center, and some private education institutions. Therefore, more Maldives students study in Malaysia, India, Sri Lanka, Britain, Australia, Egypt, and other countries.

Significantly less population of Afghanistan continues to receive higher education. At present, there are 19 public and more than 75 private higher education institutions. Meeting the demand of many students to receive higher education is the biggest challenge in education in Afghanistan.

Overall, South Asian countries have a similar geographical environment and climate, dense population growth, low level of economic development, underdeveloped higher education, low enrollment rate, and lower female participation in higher education. Therefore, South Asian countries have a strong demand for higher education, and international cooperation in higher education is the key to solving the problem.

3 Factors Influencing the Internationalization of Higher Education

The internationalization of higher education is based on political diplomacy, personnel exchange, and cultural exchange among countries. To a large extent, it serves the economic and social development and foreign policy of each country. Therefore, higher education internationalization in South Asia depends on many factors; some of the critical factors are discussed below.

3.1 Impact of Political Relations Among Countries

In 1985, the South Asia Association for Regional Cooperation (SAARC) was established. As a regional consortium, South Asia formed a community of interests based on similar geographical conditions and historical background, which promoted exchanges and cooperation among countries, promoted member states' economic and social development, and finally achieved the goal of promoting the best use of human resources. Educational sovereignty is a part of national sovereignty. Under the South Asia regional cooperation alliance mode, it is still a long way to complete the integration of higher education in the region and the win-win and coordinated development of academic and scientific research.

3.2 Influence of Different Ethnic, Cultural Differences

South Asia has a multi-cultural feature, with more than 2000 different races. It is the birthplace of Hinduism, Buddhism, Jainism, and Sikhism. There are significant differences in customs and religious beliefs. Education is an important way of cultural communication. In the internationalization of education, cultural friction, confrontation, and other obstacles inevitably become an essential factor affecting the further development of higher education cooperation.

3.3 *Impact of the Difference in Economic Development Level*

The internationalization of higher education has a great relationship with the financial support of the government. Sufficient funds are the material guarantee and foundation of the development of higher education. Therefore, the difference between the overall level and per capita level of economic development in South Asia significantly impacts its position and discourse power in the internationalization of education.

3.4 *Influence of the Development Level of Higher Education*

One of the goals of international cooperation in higher education is to achieve a win-win situation, which requires the complementarity of higher education between the two sides. However, there are many differences in the scale, characteristics, system, policy, quality, and level of higher education in different countries, which affect the circulation direction of limited higher education resources in the region.

3.5 *Impact of Public Health Emergencies*

Since 2020, the novel coronavirus pneumonia epidemic has spread worldwide. Many people have been seriously affected due to the high population density in South Asia. The novel coronavirus pneumonia rate has increased rapidly since July 2020 and has become the fastest-growing region after the Americas. It has seriously affected the development of the economy and international exchanges. As a result, there are tremendous pressure on epidemic prevention and control, social assistance, and economic recovery in the region. Due to the low level of economy and the relatively low level of information development in South Asia, the development of online communication under the influence of an epidemic situation is also limited.

4 The Current Internationalization of Higher Education

The current internationalization of higher education shows that the cooperation among countries needs to be strengthened and deepened. Internationalization is the trend of higher education development and has become one of the development strategies of colleges and universities. With the improvement of economic level and the development of information technology, the relationship of higher education among various countries is increasingly close, and the cooperation and exchange are deepened. The first meeting of members of the world's textile universities was held on December 8, 2018. The first alliance members came from thirty-three textile universities from nineteen countries; see some glances of the event in Fig. 1. The alliance is the largest collaborative organization of textile universities globally, with the highest participation rate in the universities and, in particular, the highest number of "One Belt and One Road" countries. To a certain extent, this major event proves that the cooperation and joint development among colleges and universities with industry characteristics have become the common need and future trend of the internationalization development of colleges and universities.

Education comes first in economic development. A specific higher education participation rate is a prerequisite for achieving rapid social and economic development. The overall economic level of South Asia is not high, but the economic

Fig. 1 The first meeting of members of the world's textile universities was held at Donghua University, China

growth momentum is strong. The economic growth rate of South Asian countries can be divided into three gradients from high to low. The first gradient has the highest economic growth rate, including Bhutan, Bangladesh, and India, with an economic growth rate of about 7%. The second gradient is followed by Pakistan, Sri Lanka, and the Maldives, whose economic growth rate is between 4 and 5%. The third gradient has the worst economic growth rate, including Afghanistan and Nepal, with a 1% or less economic growth rate.

The South Asian Association for Regional Cooperation (SAARC) has reached a consensus in promoting higher education and international cooperation. All countries fully realize that backward education is the leading cause of regional socio-economic backwardness, and higher education is of great significance for knowledge-based society and economic modernization. Education cooperation is of great significance to all countries in the regional economy. It is an essential basis for enhancing understanding, identity, and win-win cooperation. It is also an essential part of the regional human resource development plan. Besides trade, investment, and other significant areas of cooperation, there are genuinely complementary advantages, mature cooperation conditions, and cooperation opportunities among global economies cooperation fields with colossal development space.

The purpose of this book is to analyze the general situation of the internationalization of higher education in South Asian countries through experience exchange, cooperative discussion, and case study, and to seek complementary, symbiotic, and win-win cooperation and development to promote the international exchange of higher education, to guide for other countries to strengthen exchanges and cooperation with higher education in the region.

5 Textiles Higher Education—The Combination of Industry, Academic, and Research

The combination of industry, academic, and research is no longer a strange topic for the textile industry. Preventing the scientific research achievements of colleges and universities from being wasted has become a topic of exploration and research in the textile industry. At present, textile industry enterprises' research and development mechanisms significantly restrict the combination of industry, academic, and research in colleges and universities. Enterprises often require schools to provide mature technology that can produce direct economic benefits, but few funds are invested in the research based on technology or application. This requires the state to provide policy support for the research of some basic disciplines to work at ease. In addition, in some schools, there is a contradiction between "high level" and "survival", coupled with the internal system, mechanism and policy, and other reasons, scientific research gradually has the tendency of miniaturization, decentralization, and simple pursuit of economic benefits, which has become a considerable difficulty in the current scientific research reform of colleges and universities. The direct participation of colleges and universities in major technological progress projects of textile enterprises is related to improving colleges and universities' overall level.

6 Conclusion

Since it can be understood that the region's higher education needs sufficient improvements, it has been suffering due to various reasons discussed in Sect. 3. This gap can be easily fulfilled by cooperation and working together of different higher institutions. Textile is one of the people's basic needs, and south Asia possesses tremendous potential in the discipline. This discipline can exist only with the collaboration between academia and industry, as highlighted in Sect. 5. Thus, to get the true benefit of this discipline, it is further essential to collaborate between the region's textile-related educational institutes. The mutual collaboration of textile-related higher educational institutes of south Asia will bring a tremendous increase in the economy and development of each country along with peace and harmony. The regions' current political, environmental, and geographical conditions also support textile manufacturing and trade. Besides, the south Asian people possess the required skills for textile and clothing; the region has a long history in the trade and manufacturing of textile, as elaborated in other chapters. Thus, we believe that choosing this as a discipline would be one of the best choices for the local people. In addition, textile higher education is hoped to bring significant change in the region by internationalization higher education.

References

1. Nascimbeni F, Burgos D, Spina E, Simonette MJ (2020) Patterns for higher education international cooperation fostered by Open Educational Resources. Innov Educ Teach Int 1–11. https://doi.org/10.1080/14703297.2020.1733045
2. Wen W, Hu D, Hao J (2018) International students' experiences in China: does the planned reverse mobility work? Int J Educ Dev 61:204–212. https://doi.org/10.1016/j.ijedudev.2017.03.004
3. Abdulai M, Roosalu T, Wagoner B (2021) International Students mobility and value added through internationalization of higher education in Estonia and Denmark. Lang Intercult Commun 1–16. https://doi.org/10.1080/14708477.2021.1900214
4. Hsieh C-C (2020) Internationalization of higher education in the crucible: linking national identity and policy in the age of globalization. Int J Educ Dev 78:102245. https://doi.org/10.1016/j.ijedudev.2020.102245
5. Sahoo P, Dash RK (2012) Economic growth in South Asia: role of infrastructure. J Int Trade Econ Dev 21:217–252
6. Sahoo P, Nataraj G, Dash RK (2014) Foreign direct investment in South Asia. In: Policy, impact, determinants and challenges. Springer, Berlin
7. Khan SM (1999) South Asian association for regional cooperation. J Asian Econ 10:489–495
8. Szmigiera M (2021) Estimated number of universities worldwide as of July 2020, by country. Statista, 19 May 2021

9. Statista (2020) Number of colleges across India from 2015 to 2019. Statista Research Department, 16 Dec 2020
10. Khan N, Amin N, Kakli MB, Piracha ZF, Zia MA (2017) Pakistan Education Statistics 2015–16. National Education Management Information System, Ministry of Federal Education and Professional Training, Government of Pakistan Islamabad, Pakistan, Feb 2017
11. Behlol MG, Dad H (2020) Educational migration as a brain drain: curse or boon for Pakistani Universities. J Res Soc Sci 8:35–47

Dr. Xinfeng Yan received his Master's degree in Management from Donghua University, China, in 2007. After graduation, he worked at the International Cultural Exchange School of Donghua University. He has long-term working experience in international education and was selected as "Outstanding Manager in Charge of Overseas Student". He has been actively conducting research and has gained fruitful results on overseas student management and university internationalization. In 2014, he completed his doctoral degree in Management from Donghua University, China. Later, he was appointed as invited researcher and deputy director of the "Belt and Road Initiative" international cooperation development center of Donghua University. In 2018, he was appointed as the invited researcher and deputy director of the Institute of International Education of Donghua University. In 2019, he was a Visiting Professor at the Ss Cyril and Methodius University of Macedonia.

At present, Dr. Yan is engaged in the management and research of university internationalization, cultural exchange, and university brand image communication at Donghua University. In 2020, his research was funded by the 13th five-year Plan of China National Education Science and was also funded by Basic Operating Outlay for Chinese National Universities. He has published more than 30 papers. He has also held four international academic conferences for overseas students and published four conference proceedings as chief editor. He has taught more than ten courses for Chinese students and international students. He has been awarded the title of "Best Teacher" by overseas students of Donghua University.

Dr. Lihong Chen received her Bachelor of Science in Clothing and Textile Science at Textile Institute from Inner Mongolia University of Technology, China, in 2006. In 2009, she received her Master of Science degree in Fashion Design and Engineering Institute from Donghua University, China (China Textile University). Her research interests are the design and production of clothing, the textile industry, textile safety, and international trade market access. In 2013, she received her Doctoral of Philosophy degree in Fashion Design and Engineering Institute from Donghua University, China. She has been researching developing biofibers from agricultural products and byproducts for textile and composite applications as a visiting scholar at Textile Science Institute from the University of Nebraska-Lincoln, the United States, from September 2010 to September 2012. She has been working on the textile and apparel industry's sustainability as a postdoc in Management Engineering at the Glorious Sun School of Business and Management Institute from Donghua University, China, from January 2014 to December 2015.

Dr. Chen is an associate professor at the Shanghai International Fashion Innovation Center in Donghua University. She is engaged in the teaching and research of Clothing and Textile Science. She is a specialist and scholar within Apparel Industry Economics and Management and a researcher at Fashion and Creative Economy research center at Donghua University. She is actively involved in fashion design and engineering research and development, fashion brand marketing, communication, and the sustainable textile and apparel industry. She has completed two provincial and ministerial level projects, five individual research project works. She has published more than 40 papers and one book. Also, she has published one national standard as the second author.

Dr. Hafeezullah Memon received his B.E. in Textile Engineering from Mehran University of Engineering and Technology, Jamshoro, Pakistan, in 2012. He served at Sapphire Textile Mills as Assistant Spinning Manager for more than one year while earning his Master's in Business administration from the University of Sindh, Pakistan. He completed his Master's in Textile Science and Engineering from Zhejiang Sci-Tech University, China, and a Ph.D. degree in Textile Engineering from Donghua University in 2016 and 2020.

Dr. Memon focuses on natural fibers' research and spinning, woven fabrics, dyeing and finishing, carbon fiber reinforced composites, recyclable, and smart textile composites. His recent research interests include natural fiber-reinforced composites, textiles and management, textile fashion, and the apparel industry. Since 2014, Dr. Memon has published more than 40 peer-reviewed technical papers in international journals and conferences, and he has been working on more than ten industrial projects.

Dr. Memon is a full professional member of the society for the Advancement of Material and Process Engineering

(SAMPE), Society of Wood Science and Technology (SWST), and International Textile and Apparel Association (ITAA). Moreover, he is a registered Engineer of the Pakistan Engineering Council. He has served as a reviewer of several international journals and has reviewed more than 400 papers. Dr. Memon is a recipient of several national and international awards. Currently, he is serving as a postdoc fellow at Zhejiang Sci-Tech University. Moreover, he is elected as Chairman of the Alumni Association of International Students at Zhejiang Sci-Tech University.

Textile Academics in India—An Overview

Sudev Dutta and Payal Bansal

Abstract Textile is one of the necessities of human life. Apart from fulfilling the basic requirements of humankind, there have been significant innovations over the years in various interdisciplinary sectors such as the medical sector, geotextiles, gaseous filtration, agriculture, etc., where textile-related products have successfully been used as a replacement of other materials due to their specific features. In view of that, the present chapter briefly provides an insight into the present scenario in India as far as the academics and research-related activities in the field of textile are concerned. In addition, universities and research institutes are monitored by various agencies viz. state government, central government, private and deemed and their mode of functioning has been reported. Overall, the chapter will provide a holistic view to the reader regarding India's contribution at the global level as far as the development of the textile sector is concerned.

Keywords Textiles · Higher education · Internationalization

1 General Introduction of Textile in India

India has been renowned in the field of textiles for centuries, as the evolution of textiles has been one of the oldest at the global level. Since the 5th millennium BC, homespun cotton was used to weave the fabrics' garments, and coloring was accomplished through indigo dyes. It is also reported that the earliest surviving Indian cotton thread date to around 4000 BC, whereas the dyed fabrics are documented as far back as 2500 BC. India's textiles are embedded in every aspect of its identity and were so central abroad that it was considered the shorthand for "cotton". The ancient Romans called India's finest kinds of cotton "woven winds" because of their airy lightness. Courtly splendor was proclaimed by magnificent fabrics, and religious worship stills expression through textile substrate [1, 2]. India's rich natural resources for making and decorating textiles are unrivaled. The country's geographical regions

S. Dutta (✉) · P. Bansal
Assistant Professor, Bannari Amman Institute of Technology, Sathyamangalam, Erode, Tamil Nadu 638401, India

X. Yan et al. (eds.), *Textile and Fashion Education Internationalization*, Textile Science and Clothing Technology, https://doi.org/10.1007/978-981-16-8854-6_2

and climates provide abundant plant fibers, natural dyes, and other related raw materials for the cultivators, weavers, dyers, printers, and embroiderers to carry out their work flawlessly. In the current era, cotton and silk are the raw materials most associated with Indian textiles, and it has been the leading supplier of cotton at a global level. The country's cotton fabrics range from the sheerest muslin to robust pieces for everyday use, though fine kinds of cotton were much sought after and are a crucial element of India's textile supremacy. It is also a significant producer of a versatile range of hand-made fabrics for domestic and commercial applications. Furthermore, India's wild and semi-domesticated silks continue to provide a wide variety of yarn and fabric. Other fibers such as animal fibers extracted from sheep's wool and yak- and goat-hair are used commercially [3–5].

The Indian textile sector provides various scope for employment, including skilled and unskilled sections of society. It accounts for the employment of more than 35 million people. It may also be added that India has numerous clusters which beautifully weave and present the most attractive fabrics [6–8]. Listed here is some of the widely popular traditional fabrics/prints/embroidery:

1. **Khadi**—In India, Khadi, also known as khaddar from the British era, is a famous handwoven natural fiber cloth. During its manufacturing, fibers are spun into yarn on a spinning wheel called a charkha. It is a versatile fabric, exhibiting a cooling effect during the summer season and warm in winter.
2. **Kalamkari**—Kalamkari is another type of hand-painted or block-printed cotton textile. There are two distinctive styles of kalamkari art in India, viz. the Srikalahasti style and the Machilipatnam style. The Srikalahasti style of kalamkari is done through a freehand drawing of patterns and filling in the colors. However, the Machilipatnam style of Kalamkari work involves vegetable-dyed block-painting of fabric.
3. **Banarsi Silk Saree**—Another famous traditional widely famous has been the Banarsi silk saree known for gold and silver brocade or zari, fine silk, and opulent embroidery. The saree is decorated with an intricate design inspired by Mughal, floral, foliate motifs, mango leaves motifs and minakari work.
4. **Chikankari**—Chikankari is an art, a long-running or darning stitch worked with six strands on the right side of the fabric taken over four threads and picking up one. It results in the transformation of the plainest cotton and organdie into flowing yards of magic. The word 'Chikan' steps from a Persian word derived from 'Chic', which refers to the 'Jali' work done on marble or wood.
5. **Sambalpuri**—Sambalpuri is a traditional handwoven ikat wherein warps and wefts are tie-dyed before weaving. As a result, the fabric incorporates traditional motifs like shankha (shell), chakra (wheel), phula (flower). In this technique, the threads are first tie-dyed and later woven into a fabric, with the entire process taking many weeks.
6. **Chanderi**—Another Indian antique produced through the blend of Cotton, Silk, and Zari referred to as the Chanderi fabric. Traditional looms are practiced as a primary means of its production. These include Dobby and Jacquard

Looms. Most Chanderis have a rich gold border and butis all over the fabric. Some have gold checks or little motifs (known as butis). The yarn used in Chanderi fabric is of high quality and extra fine. Because of the non-degumming of the raw yarn, the finished fabric produced is exceptionally transparent and which, in consequence, result in sheer texture.

7. **Paithani**—Another art of weaving widely known as Paithani is about 2000 years old. This fabric variant is silk with an ornamented zari pallav, and the border looks identical on both sides. Generally, traditional vines, flowers, shapes of fruits, and stylized birds, especially peacocks, are used as motifs woven in rainbow colors.
8. **Patola**—A double ikat woven fabric known as patola, usually made from silk. Patola means "Queen of Silks," woven with outstanding clarity and precision. However, manufacturing requires labor intensiveness, and each patola article takes a long time, i.e., from six months to one year, to prepare.
9. **Pashmina Shawls**—The Pashmina shawls are made from goat's wool, having intricate embroidery. Pashmina refers to soft gold in Kashmiri. Some of these designs are hand-block printed, and those blocks sometimes date back to more than 100 years. Hand embroidery is done on the shawl, which takes more time to make it an end product. A continuous tedious effort required for construction makes it one of the costliest fabrics.
10. **Kullu Shawls**—A Kullu shawl is a type of shawl made in Kullu, Himachal Pradesh, India, featuring various geometrical patterns and bright colors. Originally, indigenous Kullvi people would weave plain shawls, but following the arrival of craftspeople from Bushahr in the early 1940s, the trend of more patterned shawls came to rise. The body of each shawl is a simple twill weave made with a colorful inlay pattern set using hand manipulations with colored woolen threads. The pattern on woolens is usually geometrical, with the color palette ranging from blue, saffron to purples and greens.
11. **Phulkari**—Phulkari, meaning 'flower work with the simplest of tools viz. a needle, a silk thread and a high degree of skill, the Punjabi's can create the most amazing flowery surface designs. Phulkari's pattern revolves around a solitary stitch.
12. **Bandhani**—Dyed in tie and dye style, the art of bandhani is a highly skilled process. Before dying, the fabric is tied into small points with threads. The knotted parts remain uncolored. Different types of tying methods are Leheriya, Mothda, Ekdali, Trikundi, Chaubandi, etc. It is also known as Bandhej and is made on superfine cotton, mulmul, or muslin.
13. **Kanjivaram Sarees**—Kanjivaram is woven from pure mulberry silk with three-ply, using thick zari forming unique and intricate designs in the sari. Weavers use the Korvai method, in which different colored yarn for body and border are interlinked.
14. **Mysore Silk Sarees**—Made from Mulberry silk and carrying historical importance, the Mysore Silk sarees are luxuriant, uniquely minimalistic, elegant, and continue to stand out with all these qualities. The sarees have the plain single

color base fabric of 100% pure silk blended with a narrow strip of gold zari border at both ends. The weaving process involves two types of looms, viz. the Dobby and Jacquard.

15. **Jamdani**—Jamdani means a vase of flowers. The history of this fabric dates back to the early ages, where its mentions can be seen scripted in Arthashashtra, written by Kautilya. The glory of this fabric can also be seen mentioned in the writings of Chinese, Italian, and Arab travelers. This fabric is constructed from unbleached cotton yarn. Saris are woven using this method are called Terchha. The part of the sari which goes over the shoulder is decorated with motifs called Jhalar.
16. **Muga silk**—Muga silk, mainly produced by the Garo community of Assam, is obtained from semi-domesticated multivoltine silkworms. It is majorly famous for its bright golden hue.
17. **Kasavu**—Kasavu is a handwoven cream-colored saree with a golden border.
18. **Madrasi Checks**—Madrasi checks are lightweight cotton fabrics with patterned texture and plaid design. Mainly used as an application of summer clothing, the fabric inherited its name from Chennai, formerly known as "Madras". Its primary color is red-dyed at the yarn stage.
19. **Bhagalpuri Silk**—Bhagalpuri silk **is** also known as the "Queen of all fabrics". The yarns are weaved with silk threads of multi-colors, which are reared from the Tussar cocoons. They are weaved into exotic and symbolic designs and motifs or plain weave in their pure texture and feel.
20. **Kota Doria Fabric**—Kota Doria fabric has a square weave pattern, making it one of the most delicate open weave fabrics. The cotton, silk, and zari (fine metal threads) yarns are weaved on the pit loom for its manufacturing. The cotton yarn provides stiffness, and silk provides luster to the fabric.
21. **Ajrak**—The term "Ajrak", derived from "Azrak", meaning blue in Arabic, as blue happens to be one of the primary colors in Ajrak printing. Printing is done with natural dyes, using carved wooden blocks for the pattern.
22. **Narayan Peth Saree**—Narayan Peth Saree is a traditional Maharashtrian saree, beautifully woven in silk with a contrasting zari border, generally with 'rudraksha' motifs. Narayanpet sarees have a checked surface design with embroidery, and the border or pallu have intricate ethnic designs. The borders and pallu of the Narayanpet Silk sari are given a contrasted look with small zari designs. Its weaving is a unique process where eight sarees are constructed simultaneously on a single loom.
23. **Kantha**—Kantha is done on tussar silk with a simple running stitch along the edges. Motifs found in Kantha embroidery include many symbols derived from ancient art and are reflective of nature, such as the sun, the tree of life, and the universe.
24. **Kunbi**—Kunbi is a chequered cotton sari in red and white with a sturdy weave good enough to be worn for farming. It has a dobby border, which is essentially a silken flat inset. Recently, designer Wendel Rodricks promoted the fabric in a fashion week.

25. **Bagru**—Bagru is a traditional printing technique using natural dyes and colors prominently dyed through the indigo dye. Motifs are transferred with wooden blocks following two styles, direct and resist style. The fabric is considered an eco-friendly one.
26. **Ilkal**—With its origin in the Ilkal town of Karnataka, it is woven using cotton warp on the body and art silk warp for the border, while silk warp for pallu portion of the saree. Pallu and body are joined with the kondi technique. Red and white are their primary colors.
27. **Sanganeri**—Sanganeri printing gained high popularity in the sixteenth and seventeenth centuries in all European countries with its Calico prints and became one of the major exports of the East India Company. Sanganer motifs are mostly floral-based and have intricate detailing. It consists of a dense large pallav and border in bright color with geometrical motifs woven in a third color. The appearance of the sari has a tribal tinge in it. The sari is usually dyed to attain the red, black, and white background colors.

2 Textile Education and Its Sub-branches in India

Textiles have become such an integral part of human's daily life that their know-how becomes a priority. Knowledge of textiles will make it easy to develop the ability to distinguish quality in fabrics and, in turn, appreciate the proper uses for the different qualities [9–11]. In India, textile education has been categorized into four forms viz. Textile Engineering, Textile Technology, Technical Chemical Processing, and Fashion Technology. The aspirants pursue their careers in all the categories mentioned above throughout the country. The categories have been elaborated on in the following sections.

2.1 Textile Engineering

Education on textile engineering deals with the study of various textile manufacturing processes and analyzing the mechanism (gearing, motion, and power transmission) of the machines used during manufacturing and value addition of different textile products. It may also be added that this field of study includes production and mechanical-related aspects of the textile sector. It combines the principles of engineering with specific knowledge of textile equipment and processes. Many universities and colleges have launched specialization courses in textile engineering in India. The syllabus of Textile Engineering allows students to know about the interaction of the material with the machine, energy conservation, natural and man-made materials, pollution, waste control, and safety and health. The core subjects within Textile Engineering are fiber, yarn and fabric formation, textile testing, chemical processing, design and structure of fabric, and garment technology [5]. Textile Engineers are usually recruited by top textile plants and companies in India as well as abroad.

This category focuses on designing and controlling the fiber, machinery, products, and apparel and textile process. This engineering field draws inputs from various other branches of engineering like chemical, instrumentation, computer, structural, electronic, and mechanical engineering. It refers to all the activities and methods involved in the process of textile manufacturing. Further, it is also added that textile engineering consists of the laws, principles, and scientific techniques utilized to develop and manufacture textile material. Therefore, this category also covers the principles of science that analyze the polymers involved in the formation of textile fiber. The textile industry involves manufacturing different types of substrates. Apart from the basic manufacturing processes, i.e., spinning, weaving, bleaching, dyeing, printing and finishing, the industry has witnessed unprecedented growth in fashion design and garment manufacturing in the recent past. Hence, this engineering category also provides light of knowledge about fashion and garment-related aspects.

Furthermore, this education class is also included in this education class due to the increasing demand for textiles in various application areas such as medicine, research areas like space, etc. Textiles are seen everywhere: clothes, bed sheets, draperies, carpeting, upholstery fabrics, or towels. The science behind the production of all these things is textile engineering. Textile engineers design and develop the processes, equipment, and procedures that create these fibers, fabrics, and Yarns. Textile Engineering demands a lot of research, creativity, and innovation, considering the demand for fashionable attire.

Further, it is added that one of the most emerging topics in the Indian textile education sector under textile engineering has been technical textiles. Textile products manufactured for non-aesthetic purposes can be categorized under technical textile, where function is the primary criterion. Technical textiles are textile materials and products manufactured primarily for their technical performance and functional properties rather than aesthetic and decorative characteristics. These include textiles for automotive applications, medical textiles (e.g., implants), geotextiles (reinforcement of embankments), agro textiles (textiles for crop protection), and protective clothing (e.g., heat and radiation protection for firefighter clothing, molten metal protection for welders, stab protection and bulletproof vests, and spacesuits). The global growth rate of technical textiles is about 4% per year. Technical textiles can be divided into many categories, depending on their end-use. The classification system developed by Techtextil, Messe Frankfurt Exhibition GmbH, is widely used in Europe, North America, and Asia. Techtextil specifies 12 application areas: Agrotech, Buildtech, Clothtech, Geotech, Hometech, Indutech, Medtech, Mobiltech, Oekotech, Packtech, Protech, and Sporttech, Architect (Architectural textile), Military textiles, Autotech (Automobile textiles), Smartech (Smart Textiles), Wearable computers. These are sometimes spelled Agrotex, Buildtex, Clothes, Geotex, Hometex, Indutex, Medtex, Mobiltex, Oekotex (Ecotex), Packtex, Protex, and Sportex.

Textiles are utilized for innumerable purposes other than the manufacturing of garments. They are used to manufacture carpets and furnishings like bedsheets and bed covers, quilts, tablecloths, curtains, towels, etc. It is also used for rags and dusters, tents and nets, kites, and parachutes. Textiles have industrial uses as well. According to its usage and texture, textiles are classified as Technical Textiles—made of zinc oxide and nanowires, which enable protection from heat and radiation. Examples of Technical Textiles are protective clothing like firefighting suits, those designed especially for astronauts. They are incorporated in firefighting suits, shuttle space suits, carbon fiber composite rocket exhausts, and even nose cones for NASA. Medical Textiles—besides gauze and bandage, textiles are also used as sutures, orthopedic implants, vascular grafts, artificial ligaments, artificial tendons, heart valves, and even artificial skins. Recent advances in medical textiles have led them to be used as extracorporeal devices, including artificial kidneys, artificial liver, or mechanical lungs. New materials find specialized applications like antimicrobial and antifungal fibers, additives used in barrier fabrics, post-operative abdominal binders, applications in neurodermatitis treatment, and various other wound management and surgical treatments. Agro textiles protect the farmers and Agricultural Engineers from the adverse effects of soil and pollution, and Geo-textiles—which protect the people during construction work [12, 13].

2.2 *Textile Technology*

Textile Technology is a division of engineering where the principles of scientific techniques are utilized for manufacturing and development in all types of yarns and textile fabric from textile fibers. It also deals with the study of chemical and physical principles, which are utilized to analyze the behavior of polymers involved in the formation of textile fiber. The significant role of this category is to design and control all aspects of textile and apparel processes, types of machinery, products, and fibers. The textile sector is playing a significant role in the growth of the Indian economy. In India, there are two important textile sectors such as handloom sector and the mechanized sector. These both have full growth potential as there are thousands of textiles mills all over the country. The core subjects within Textile Technology are Technical Textiles, Knitting and Knit, Weaving, Colouration Technology, Yarn and Non-woven Technology, Textile Materials, and Performance Evaluation.

Textile technology applies scientific and engineering principles to designing and controlling all aspects of fiber, textile, and apparel processes, its products, and machinery. These include natural and man-made materials, the interaction of materials with machines, safety and health, energy conservation, and waste and pollution control. There is broad scope for research in this area as the industry demands the need to improve currently available products and develop new ones [12, 14].

2.3 Textile Chemical and Processing

The Indian textile sector has been keenly focusing on the value addition aspect of textile products. Therefore, a separate course about dyeing and processing is being carried out at various textile educational institutes in the country. The study deals with various processes and mechanisms involved during dyeing and processing. Besides, the education sector keeps itself updated with various innovations in the said domain [15].

2.4 Fashion Technology

Fashion technology is a term that describes innovative technologies in material procurement and fashion design and their applicability in fashion manufacturing, transportation, and retail. In simple terms, fashion technology creates contemporary tools for the fashion industry, intending to improve our production and consumption of fashion. Modern tools such as the use of AI in fashion design, 3D printers instead of sewing machines, lab-made leather alternatives instead of animal leather, body scanners instead of measuring tape, augmented reality, and VR for retail, and much more. Seeing the popularity of fashion and its trends, many colleges started a specialized engineering course in 'Fashion Technology' to impart deep knowledge of the industry. Students also showed a great interest in the course, and now, it has become one of the popular courses among students. The course teaches the basics of the fashion industry and sharpens the students' creativity skills, which is the core element of the course. India's big fashion tycoons like Rohit Bal, Sabyasachi Mukherji, Manish Malhotra have already set a benchmark in this industry, and many youngsters follow their steps. Subjects of fashion technology: Textile and Apparel Design Leather Technology Fashion Merchandising Visual Merchandising Footwear Design Jewellery/Gem Designing [16].

3 Textile Courses Are Offered in Different Institutes/Universities/Colleges

3.1 Undergraduate Program

It is crucial to find the right program in terms of affordability, length of study, and physical location in making a wise choice for a fulfilling educational journey. After optimizing the aspects mentioned earlier, the undergraduate program in India offers three years diploma and four years degree courses to the undergoing aspirants after successful completion of their tenure. It may also be noted that the course is carried based on the semester system; each semester is of six months. The program's first year comprises an overview of essential engineering branches such as mechanical, electrical, and electronics. Besides, studies about applied and general sciences are also covered during the first year of the course.

Further, the candidates are made aware of the core field (textiles) in the second year. The third and final year deals with studying the advanced version of textiles, such as using technology to develop new materials are other particular focus areas in a broad-based education in textile studies. Further, it is also added that at the end of the second and third year, the candidates undergo four months of training in renowned textile industries, where they are given exposure to the industry's work culture. In the final year of the program, the candidates are assigned experimental projects related to industrial problems.

A degree in textile widens the doors to professional careers in the core field and fashion design and merchandising of apparel. Retail buyers possess textile degrees and attain positions in museums as textile conservators or curators of collections. Graphic designers and printmakers are trained in textile programs, and many graduates seek advanced studies and become lecturers and post-secondary instructors. Overall, an apparel and textiles bachelor's degree program focus on the intersection between apparel design and business [17].

3.2 Postgraduate Program

The postgraduate program offers a two-year Master of Technology Course in various textiles (Engineering, Technology, and Processing). Classes vary according to the area of specialization and interest of the candidates. Some possible classes in which the candidates pursue their project studies include biomaterials, composites, product development, comfort and structures, medical textiles, filtration textiles, agro textile, dyeing, and processing [17]. It may also be added that before the candidates choose an area of concentration for undergoing experimental projects, they must complete theoretical courses that introduce various fiber, yarn, and fabric-based processes and products. Besides the course also covers subjects related to dyeing, finance, management as well as marketing.

3.3 Doctoral and Post-Doctoral Program

Generally, the candidates pursuing their career in academics and research are inclined to pursue Ph.D. and post-doctoral research in specialized domains related to textiles and the interdisciplinary fields. Therefore, the candidates qualifying for the national level entrance test to Ph.D. benefit from a five-year scholarship during the doctoral program and three years in the post-doctoral program. During the Ph.D. program, a candidate is considered Junior Research Fellow (JRF) for the first two years and Senior Research Fellow (SRF) for the final years. In the first year of the program, the candidate has to clear the course work comprising four subjects, with Research Methodology being the compulsory subject. The remaining three subjects chosen by the candidates are usually relevant to their research topic [17]. The details of them have been given in the next section.

4 Academic Institutes

Numerous academic institutes in India produce many technocrats in the field of textiles every year. India's contribution to the world textile arena has been quite significant over the years, and one of the primary reasons for this has been the time to time up-gradation of textile education in the country. Innovative methods and sustainable developments are being opted by the textile education system. Textile industries and education sectors are working together on various innovative research projects. There have been many collaborations and MOUs signed by different textile institutes with the industrial sectors; as a result of that, people are becoming more knowledgeable and aware of textiles over time. There are different subjects of textiles, viz. wet process, Textile chemistry, garments manufacturing, fabric manufacturing, spinning, or yarn manufacturing. Besides, now a day's technical textile has become one of the most prevalent areas of study in India, as the textile materials are incorporated in a number of interdisciplinary applications such as civil and constructions to prevent soil erosion, aerosol filtration, for efficient capturing of particulate matters, medical sector, for an artificial kidney transplant, sutures and many more. The institutes offer different courses, as already mentioned in the previous section. The institutes offering the courses as mentioned earlier are monitored by a governing body known as AICTE (All India Council for Technical Education), which was set up in 1945 as a national-level Apex Advisory Body to conduct a survey of the facilities available for technical education and to promote development in the country in an integrated manner [18–21].

4.1 Central Institutes

Central institutes are under the purview of the Department of Higher Education in the Ministry of Education (MHRD). In general, institutes in India are recognized by the University Grants Commission (UGC), which draws its power from the University Grants Commission Act, 1956. In addition, 15 Professional Councils are established, controlling different aspects of accreditation and coordination. These institutes are funded and run by the Central government. There are six institutes in the central sector in India and four in the State sector of the Indian Institutes of Handloom Technology (IIHTs). All IIHTs provide a three-year Diploma in Handloom and Textile Technology. Central sector IIHTs are under the administrative control of the Ministry of Textiles, the Government of India. The Indian Institute of Technology

Delhi, National Institute of Technology Jalandhar, National Institute of Fashion Technology also come under the center institutes. These institutes have been categorized as institutes of national importance by the Indian government. The courses offered by institutions, as mentioned earlier, are given in Table 1.

4.2 *Deemed Institutes*

Deemed institutes are those institutes that have been granted a status of autonomy by the recommendation of UGC because of the high standard of work carried out there. These are granted by the Ministry of human resources department on the advisory of UGC. An Institution of Higher Education, other than universities, working at a very high standard in a specific area of study, can be declared by the Central Government on the advice of the University Grants Commission (UGC), as an Institution' Deemed-to-be-university'. Institutions that are 'deemed university' enjoy the academic status and privileges of a university. Table 2 shows the institutes that come under deemed institutes.

4.3 *State Institutes*

State universities come under the state government though they also require recognition by UGC. These are the university maintained and administered by one of the states as part of the state public education system. State institutes are funded by the governments of the respective states and are public varsities. These institutes are under the administrative control of respective state governments and bodies. Table 3 shows all the detail of state institutes in India.

4.4 *Private Institutes*

The institutes monitored by AICTE and having a major share of funding through private sources comes under the private category. Table 4 illustrates the details of some details of such private institutes where textile-related courses are being offered.

5 Textile Research Associations

The main objective of these research Associations is to carry out research and render consultancy services to the textile industry on various aspects of textile technology to reduce the cost and improve the quality and durability of fabrics, reducing pollution,

Table 1 Central Textile Institution in India

Name	Course offered	Year of establishment
Indian Institute of Handloom Technology, Varanasi	• Diploma in Handloom and Textile Technology • Post Diploma in Textile Processing	1956
Indian Institute of Handloom Technology, Salem	• Diploma in Handloom and Textile Technology • B.Tech. Degree in Handloom and Textile Technology	1960
Indian Institute of Handloom Technology, Guwahati	• Diploma in Handloom and Textile Technology	1982
Indian Institute of Handloom Technology, Jodhpur	• Diploma in Handloom and Textile Technology	1993
Indian Institute of Handloom Technology, Bargarh	• Diploma in Handloom and Textile Technology	2008
Indian Institute of Handloom Technology, Fulia	• Diploma in Handloom and Textile Technology	2015
Northern India Textile Research Association (NITRA)	• B.Tech. in Textile Engineering	1974
Indian Institute of Technology Delhi (IIT Delhi)	• B.Tech. in Textile technology • M.Tech. in Textile Engineering • M.Tech. in Fiber Science and Technology • M.Tech. in Textile Chemical Processing • Ph.D. in Textile Technology • Post Doc	1961
Dr. Br Ambedkar National institute of technology (NIT Jalandhar)	• B.Tech. in Textile Technology • M.Tech. in Textile Engineering and Management • Ph.D. in Textile Technology	1987
National Institute of Fashion Technology (NIFT)	• B.Des. (Fashion Design) • B.Des. (Leather Design) • B.Des. (Accessory Design) • B.Des. (Textile Design) • B.Des. (Knitwear Design) • B.Des. (Fashion Communication) • B.F.Tech. (Apparel Production) • M.Des. (Master of Design) • M.F.M. (Master of Fashion Management) • M.F.Tech. (Master of Fashion Technology)	1986

Table 2 Deemed Textile Institution in India

Name	Course offered	Year of establishment
Dayalbagh Educational Institute (DEI) Technical College, Agra	• Diploma in Garment Technology • Diploma Textile Design (DTD)	1927
Vignan's Foundation for Science, Technology and Research	• B.Tech. in Textile Technology	1997
Arunachal University of Studies	• B.Sc. in Textile Design	2012
Himalayan University, Itanagar	• B.Tech. Textile Engineering	1956
Assam Textile Institute	• Diploma Textile Technology, Garment Technology and Fashion Technology	1920
Institute of Chemical Technology—ICT	• B.Tech. (Fibers and Textiles Processing Technology)	1933
Institute of Chemical Technology, Mumbai	• B.Tech. Fibers and Textiles Processing Technology	1933

conserving energy, and utilizing waste, adopting new technology and improving the technology in the decentralized textile sector [22]. Some of the leading research centers of the country have been listed in Table 5.

6 Textile Events Specific to Academia

To improve the academic interest and enrich professional interaction, the premier textile institutes and universities in India organizes numerous events such as conferences, symposium, workshops, and short-term courses every year so that the aspirants remain updated to the current textile scenario. The conferences are categorized into two sections, viz. international and national. The eminent personalities and delegates from abroad present the know-how in their expertise area in the international conference. However, in national conferences, the domain remains specific within the country. Similarly, the other events, viz. symposiums, workshops, and short-term courses, also aim to enhance the textile knowledge of the candidates through lectures of technocrats serving in leading academic institutes and organizations.

7 Schemes Under Textile Ministry

The Government of India has many achievements made in the textiles sector, and some of the significant achievements and contributions are given below [23]:

- Eight Apparel and Garment making Centers set up in all NER States and Sikkim for promoting garment manufacturing in NER.

Table 3 State Textile Institution in India

Name	Course offered	Year of establishment
Giani Zail Singh Campus College of Engineering and Technology (Maharaja Ranjit Singh PunjabTechnical university)	• B.Tech. in Textile Engineering • M.Tech. in Textile Engineering (Part-time) • Ph.D. in Textile Engineering (Part-time)	1989
Punjab Agricultural University	• M.Sc. in Apparel and Textile Science • M.Sc. in Fashion Designing	1962
Chaudhary Sarwan Kumar Himachal Pradesh Krishi Vishvavidyalaya	• M.Sc. in Textile and Apparel Designing	1978
Jawaharlal Nehru Government Engineering College, Sundernagar, Himachal Pradesh	• B.Tech. in Textile Engineering	2006
The Technological Institute of Textile and Sciences (TIT&S)	• B.Tech. in textile technology • Bachelor of Technology in Fashion and Apparel Engineering • Bachelor of Technology in Textile Chemistry • Bachelor of Technology in Textile Technology • Bachelor of Technology in Fashion and Apparel Engineering • Bachelor of Technology in Textile Chemistry • Bachelor of Technology in Textile Technology • Master of Fashion and Apparel • Master of Technology in Textile	1937
PIET—Panipat Institute of Engineering	• B.Tech. in Textile Engineering	2006
Govt Polytechnic, Hisar	• Diploma Textile Design (DTD) • Diploma Textile Processing (DTP) • Diploma Textile Technology (DTT)	1992
Shivaji University	• M.Tech. (Textile Chemistry) • B.Tech. (Textile Chemistry) • B.Tech. (Textile Engineering	1962
Indian Institute of Carpet Technology IICT, Bhadohi	• Bachelor of Technology (BTech Textile Technology)	2001
Government Girls Polytechnic, Allahabad	• Diploma Textile Design (DTD)	1992

(continued)

Table 3 (continued)

Name	Course offered	Year of establishment
Uttar Pradesh Textile Technology Institute Kanpur	• Bachelor of Technology in Textile Technology • Bachelor of Technology in Textile Engg. • Bachelor of Technology in Textile chemistry • Bachelor of Technology in Man-made fiber Technology • M.Tech. in Textile Technology and Textile chemistry	1914
Government Polytechnic College, Gulzarbagh	• Diploma in Textile Engineering	1954
Government Polytechnic College, Bhagalpur	• Diploma in Textile Engineering	1955
Government Polytechnic College, Anantapur	• Diploma in Textile Technology	2009
Residential Girls' Polytechnic College	• Diploma in Textile Chemistry and Design	1996
AAFT University of Media and Arts	• Diploma in Fashion Design • B.Sc. in Fashion Design • B.Des. in Fashion Design • M.Sc. in Fashion Design	2018
Indira Kala Sangit Vishwavidyalaya—IKSVV	• B.Voc. in Fashion Design • B.Voc. in Textile Design	1956
The Maharaja Sayajirao University of Baroda	• B.Tech. in Textile Technology • M.Tech. in Textile Technology	1881
L D College of Engineering	• B.E. in Textile Technology • M.E. (Textile Engineering)	1948
Gujarat Technological University	• B.E. (Textile Technology) • Diploma in Textile Technology	2007
Fortune Institute of Management and Technology	• B.Tech. (Textile Engineering)	2010
Bihar Institute of Silk and Textiles, Bihar Institute of Silk and Textiles	• B.Tech. in Textile engineering • B.Tech. in Silk Technology	1994
Govt Sri Krishnarajendra Silver Jubilee Technological Institute, Bangalore	• Bachelor of Engineering (BE Silk Technology) • Bachelor of Engineering (BE Textile Technology)	1938

(continued)

Table 3 (continued)

Name	Course offered	Year of establishment
K. Gujral Punjab Technical University	• B.Sc. Fashion Design Knits (FDK) • B.Sc. Textile Design (TD) • B.Sc. Fashion Design (FD) • B.Sc. (Fashion Tehnology) • Bridge Program-Fashion Design and Textile Design • Bridge Program-Fashion Design Knits • Bridge Program-Fashion Marketing Management and Garment Manufacturing Technology • Master of Design in Fashion and Textiles (M. Des. F&T) • M.Sc. Fashion Design • M.Sc. Fashion Marketing and Management (FMM) • M.Sc. Garment Manufacturing Technology (GMT) • Post Graduate Diploma in Fashion Design and Clothing Technology	1997
Institute of Textile Technology Orissa	• Diploma Course in Textile Technology	1983
Bapuji Institute of Engineering and Technology, Davangere	• Bachelor of Engineering (BE Textile Technology)	1979
RTE Societys Rural Engineering College, Hulkoti	• Bachelor of Engineering (BE Textile Technology)	1980
SLN College of Engineering, Raichur	• Bachelor of Engineering (BE Textile Technology)	1979
Sri Krishnarajendra Silver Jubilee Technological Institute, Bangalore	• Master of Technology (MTech Textile Technology) • Bachelor of Engineering (BE Silk Technology) • Bachelor of Technology (BTech Textile Technology)	1938
Skiltek Educational Institutions	• B.Tech. (Textile Engineering)	1979
Rashtrasant Tukadoji Maharaj Nagpur University—RTMNU	• B.E. (Textile Technology)	1923
Visvesvaraya Technological University—VTU	• B.E. (Textile Technology) • B.Tech. (Textile Technology) • M.Tech. (Textile Technology)	2010

(continued)

Table 3 (continued)

Name	Course offered	Year of establishment
College of Engineering Technology, Bhubaneswar	• B.Tech. in Fashion and Apparel Technology • B.Tech. in Textile Engineering • M.Tech. in Textile Engg • M.Tech. in Textile Chemical Processing	1981
University of Mumbai	• B.F.A. in Textile Design	1857
Veermata Jijabai Technological Institute, Mumbai	• B.Tech. Textile Technology	1887
Sasmira Institute of Man-Made Textiles, Mumbai	• Diploma in Textile Technology	1950
Dr. Dharmambal Government Polytechnic College for Women	• Diploma in Garment Engineering	1962
Arulmigu Palani Andavar Polytechnic College	• Diploma in Garment Engineering	2003
MLV Government Textile and Engineering College	• B.Tech. (Textile Technology) • M.Tech. (Textile Technology) • B.Tech. (Textile Engineering)	1998
Government College of Engineering and Textile Technology, Serampore	• B.Tech. in Apparel Production Management and textile technology	1908
Institute of Jute Technology, Calcutta University	• Textile Engineering Courses	1951
Government College of Engineering and Textile Technology, Berhampore	• Textile Engineering Courses	1927
West Bengal University of Technology	• Apparel and Production Management • Textile Technology (UG and PG)	2001

- Launch of Indian Handloom Brand for providing brand value for handloom products.
- Ministry working on the promotion of technical textiles.
- Integrated scheme for development of silk industry
- Scheme for capacity building in textile sector (Samarth)
- Block Level Clusters (BLCs)
- Weaver Mudra Scheme
- Handloom Weaver Mudra Portal
- Bunkar Mitra
- Handlooms promotion through e-marketing
- Group workshed scheme (GWS)
- Yarn Bank Scheme
- Scheme for integrated textile park (SITP)

Table 4 Private Textile Institution in India

Name	Course offered	Year of establishment
Dream Zone School of Creative Studies, Bilaspur	• Professional in Garment Design • Professional in Fashion CAD • Certificate in Fashion Retail • Certificate in Reach Fashion Studio • Certificate in Reach CAD • Certificate in Manual Pattern Making • Certificate in Fashion Merchandising • Diploma in Fashion CAD • Diploma in Pattern Making • Master Diploma in Fashion Design	2005
Shri Rawatpura Sarkar Universit	• Diploma in Fashion Design • Post Graduate Diploma in Fashion Design • B.Sc. in Textile Design • B.Sc. in Fashion Design	2018
Sarvajanik College of Engineering and Technology—SCET	• B.E. (Textile Technology)	1995
HKES Polytechnic, Raichur	• Diploma Textile Technology (DTT)	1986
Rural Engineering College—REC	• B.E. (Textile Technology)	1982
Anuradha Engineering College	• B.E. (Textile Technology)	1993
PSG College of Technology	• B.E. (Textile Technology) • M.E. (Textile Engineering)	1951
Bharath Post Graduate College	• B.Tech. (Textile Engineering)	2000
Shri Vaishnav Institute of Technology and Science—SVITS	• B.E. (Textile Technology)	1995
Balaji Institute of Management and Technical Studies	• B.Tech. (Textile Engineering)	2008
SSM Technical Institute	• Diploma in Textile Technology • Diploma in Textile Processing • Diploma in Textile Technology [Design and Weaving] • Diploma in Textile Technology [Man Made Fibers] • Diploma in Textile Marketing and Management • Diploma in Garment Technology	1998
Bannari Amman Institute of Technology	• B.Tech. Textile Technology • B.Tech. Fashion Technology	1996

(continued)

Table 4 (continued)

Name	Course offered	Year of establishment
SSM College of Engineering, Namakkal, Tamil Nadu	• B.Tech. (Textile Technology) • B.Tech. (Textile Chemistry) • M.Tech. (Textile Chemistry)	1988
DKTE Society's Textile Engineering Institution	• B.Tech. in Textile Technology • B.Tech. in Manmade Textile Technology • B.Tech. in Textile Plant Engineering • B.Tech. in Textile Chemistry • B.Tech. in Fashion Technology • Postgraduate program in Textile Engineering • Ph.D. Research center in Textile Engineering	1982
Chhattisgarh University / Amity University	• B.Des. (Fashion Communication) • B.Des. (Fashion Design) • (Fashion Design)—3 Continent • B.Des. (Fashion Technology) • B.Des. (Textile Design) • M.A. (Fashion Retail Management) • M.Des. (Fashion and Textiles) • M.A. (Fashion and Textile Merchandising) • M.B.A. (Fashion Management) • Ph.D. (Fashion Management) • Ph.D. (Fashion Design)	2005
Arulmurugan Polytechnic College	• Diploma in Garment Engineering	2007
BS Negi Mahila Pravidhik Prashikshan Sansthan	• Diploma in Garment Engineering	1987
GRG Polytechnic College	• Diploma in Textile Technology • Diploma in Garment Engineering	1980
Kamala Nehru Polytechnic for Women	• Diploma in Garment Engineering	
PAC Ramasamy Raja Polytechnic College	• Diploma in Textile Technology • Diploma in Garment Engineering	1963

- Integrated processing development scheme (IPDS)
- Ambedkar Hastshilp Vikas Yojana
- Design and Technology Upgradation
- Carpet weaving in Kashmir
- Export of Handicraft Products
- Visa exemption to foreign designers in handicrafts sector
- Minimum Support Prices (MSP) Operation of Seed Cotton (Kapas)

Table 5 Textile research centers

Name	Location	Year of establishment
Northern India Textile Research Association (NITRA)	Delhi	1974
The Bombay Textile Research Association (BTRA)	Mumbai	1954
South India Textile Research Association (SITRA)	Coimbatore	1956
Ahmedabad Textile Industry's Research Association (ATIRA)	Gujarat	1947
Synthetic and Art Silk Mills' Research Association	Maharashtra	1950
Man-Made Textile Research Association	Gujarat	
Indian Jute Industries' Research Association (IJIRA)	Kolkata	1937
Wool Research Association	Maharashtra	1963
Defence Research and Development Organisation—DRDO	Delhi	1958
Terminal Ballistics Research Laboratory (TBRL)	Chandigarh	1961
Central Silk Board	Karnataka	1948
Wool federation	Himachal Pradesh	1988

- Cotton Corporation of India (CCI)
- Revival of National Textile Corporation (NTC)
- Pashmina shawl from Kashmir
- Bundles of Jute Fiber in a jute mill
- Northeast Region Textile Promotion Scheme (NERTPS)
- For the silk sector, government has introduced Silk Samagra Scheme to promote R&D and brand image of Indian silk at the global level and provide support to beneficiaries from nursery to fabric production stage for improvement in productivity and quality.
- Technology Mission on Cotton (TMC)
- Technology Upgradation fund Scheme (TUFS)
- Setting up of Apparel Training and Design Centres (ATDCs).

8 Conclusions

The chapter elaborated on the present status of Indian academia and professional degrees awarded in textiles and its related areas at various levels (graduate, postgraduate, and doctoral). The education system analysis revealed that India's government has been highly inclined to uplift textile education with the various innovations and developments occurring at the global level in the said area. It may also be added that the textile ministry in India has been primarily focusing on technical textiles as it is one of the emerging areas in the global sector. Further, through a number of industrial collaborations and alliances with reputed foreign universities, the textile education system in the country remains updated. Hence, based on the analysis mentioned above, it can be envisaged that Indian textile education is on the right track. However, the scope of improvement will always be there.

References

1. https://medium.com/@noopurshalini/textiles-of-india-d9f5e5310dc6, accessed on 16 Dec 2020
2. https://en.wikipedia.org/wiki/Textile_industry_in_India, accessed on 16 Dec 2020
3. Lally J (2009) The pattern of trade in seventeenth-century Mughal India: towards an economic explanation. London School Econ Work Pap 120:22–25
4. Ray HP (2005) Far-flung fabrics—Indian textiles in ancient maritime trade. Text Indian Ocean Soc 74–75
5. Gittinger M (1982) Master dyers to the world: technique and trade in early Indian. Dyed Cotton Text 1–4
6. Appadurai A (1986) Swadeshi (home industry): cloth and Indian society. In: The social life of things: commodities in cultural perspective, pp 285–322
7. https://medium.com/@noopurshalini/textiles-of-india-d9f5e5310dc6, accessed on 18 Dec 2020
8. https://www.craftsvilla.com/blog/indian-prints-fabrics-bandhani-ikat-batik/, accessed on 23 Dec 2020
9. http://ecoursesonline.iasri.res.in/mod/page/view.php?id=31009, accessed on 23 Dec 2020
10. Corbman BP (1983) TEXTILES: fiber to fabric, 6th edn. McGraw-Hill International, pp 2–3
11. Joseph ML (1981) Introductory textile science, 4th edn. CBS College Publishing, New York, pp 1–6
12. https://www.myklassroom.com/Engineering-branches/16/Textile-Technology, accessed on 08 Jan 2021
13. https://www.technicaltextile.net/articles/technical-textiles-and-their-applications-3922, accessed on 08 Jan 2021
14. https://www.myklassroom.com/Engineering-branches/57/Textile-Technology-(SS), accessed on 08 Jan 2021
15. Uddin F (2019)Introductory chapter: textile manufacturing processes. Intech Open 1–4. https://doi.org/10.5772/intechopen.87968
16. https://www.myklassroom.com/Engineering-branches/39/Fashion-Technology, accessed on 16 Jan 2021
17. https://study.com/textile_education.html, accessed on 20 Jan 2021
18. https://targetstudy.com/colleges/btech-textile-engineering-degree-colleges-in-india.html, accessed on 20 Jan 2021
19. https://collegedunia.com/btech/textile-engineering-colleges, accessed on 20 Jan 2021
20. https://www.shiksha.com/engineering/colleges/b-tech-in-textile-engineering-colleges-india, accessed on 20 Jan 2021
21. https://www.collegedekho.com/engineering/textile_engineering-colleges-in-india/, accessed on 20 Jan 2021
22. http://texmin.nic.in/about-us/textile-research-associations, accessed on 20 Jan 2021
23. http://texmin.nic.in/schemes, accessed on 20 Jan 2021

Sudev Dutta The author is a researcher in textile and has a research experience of more than nine years in the relevant field with three years of teaching in the said area. The author completed his Bachelor of Technology in Textile Engineering in 2010. Further completed Master of Technology in 2015 and Doctoral Research in the related field in 2021. With the authorship of various research papers and articles, the author has been keen to make a vital contribution to textiles. In writing the present book chapter, the writer has consulted various books, journals, and articles relevant to dyeing and chemical processing for extracting useful information that can be beneficial for the readers.

Payal Bansal The author is a researcher in textile and has research experience of more than six years in the relevant field. The author completed her Bachelor of Technology in Textile Engineering in 2013. Further completed Master of Technology in 2015 and presently is pursuing Doctoral Research in the related field. With the authorship of various research papers, articles, and book chapters, the author has been keen to make a vital contribution to textiles. In writing the present book chapter, the writer has consulted various books, journals, and articles relevant to dyeing and chemical processing for extracting useful information that can be beneficial for the readers.

Brief Analysis on the Past, Present, and Future of Textile Education in Bangladesh

Mohammad Fahim Uddin

Abstract The scenario of textile industries in Bangladesh has been progressing rapidly for the past few decades. As the second-largest exporter for readymade garments of the world, only next to China, the government of Bangladesh focuses much attention on sustaining the current position and developing various strategies for further achievements. Since the birth of the RMG business in the country, one of the primary goals in the nation's educational policy was to build a suitable environment for textile studies and related disciplines. This study highlights the scenario of that sector from an analytical perspective, also evaluating the present situation of textile engineering among Bangladeshi universities. Moreover, it gives a glimpse of the research opportunities for interested personnel and their future career in the fields. In addition, it provides a brief idea about how the nation is communicating and co-operating with universities and institutes outside the country and why Bangladesh should be considered a good choice for textile studies among students and researchers worldwide.

Keywords Textile · Bangladesh · University · Education · Curriculum · Research

1 Introduction

For the last couple of decades, the Readymade Garment industry (RMG) has been the largest source of foreign currency in Bangladesh. It created substantial employment opportunities and has seen tremendous growth by utilizing the semi-skilled and unskilled labor force. In 2016, Bangladesh secured the second position in manufacturing garments goods and is now the world's second-largest apparel exporter of western "fast" fashion brands. In the financial year 2018–19, the national export contribution of RMG (both woven and knit) was 83.49%, or nearly USD 34.13 billion, which is also 12.26% of the GDP (Fig. 1).

M. F. Uddin (✉)
College of Material Science and Engineering, Zhejiang Sci-Tech University, Hangzhou 310018, China

X. Yan et al. (eds.), *Textile and Fashion Education Internationalization*, Textile Science and Clothing Technology, https://doi.org/10.1007/978-981-16-8854-6_3

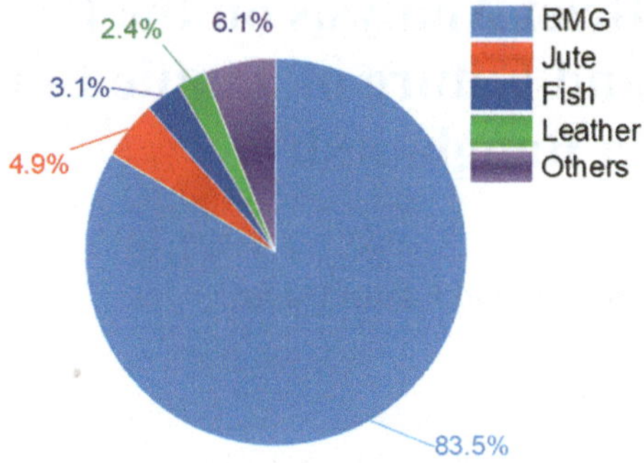

Fig. 1 Pie Chart showing the main exports of Bangladesh. *Source* Statistics Department, Bangladesh Bank

According to BGMEA, nearly 6.5 million people directly work in 4482 garment industries, and around 80% are women. Also, 15 million more people are working in support industries related to this trade. The sector also created another 0.5 million jobs in the textile-based waste recycling industry. Many factors have facilitated the development and growth of the industry, including geographical location, low-cost labor force, good quality products, and different national and international policies, which made Bangladesh one of the largest exporters of readymade garments in the world. Nevertheless, continuous research, training, development, and building skilled human capital are essential to keep pace with the world.

2 History of Textile Education in Bangladesh

The Chronology of Textile Education in Bangladesh began way back in 1911; let us have a brief look at its 100 years history [1]:

- From 1911 to 1929: Textile education officially started in Indian Subcontinent, with only nine district weaving schools and 25 peripatetic weaving institutes.
- 1929 to 1935: Some 29 Satellite-types vocational schools started a "Semi-formal educational training" program.
- 1935 to 1947: 47 monotechnic institutes started in United Bengal, providing one-year certificate courses and six months artisan courses.
- After 1947: 33 institutes started in East Pakistan, six district weaving schools and 27 peripatetic weaving institutes. Also, the Dhaka Weaving School was upgraded to be the only dedicated Textile university in the country.
- In 1950: Dhaka weaving school transferred to Tejgaon industrial area from Narinda.
- In 1954: Formal education in Textiles began as Dhaka weaving school turned into a diploma institute called East Bengal Textile Institute (later called Textile College).
- In 1958: Textile College started to deliver diploma courses in Textile Technology.

- In 1977: The "Department of Textiles" was opened for introducing Vocational and Diploma education.
- In 1978: Diploma institute became country's only college for Textile technology. Also, it was upgraded to Bachelor level (B.Sc. in Textile Technology) and was renamed the College of Textile Technology (CTT).
- In 1979: Bangladesh Textile Mills Corporation opened the Textile Industry Development Centre with technical and financial support from UNIDO, which was later upgraded to the National Institute of Textile Training Research and Design (NITTARD).
- In 1980: Two years certificate courses were introduced. 5 District Weaving Schools turned into District Textile Institutes. Another new District Textile Institute was set up.
- In 1986: National NGOs started imparting Textile Vocational training.
- In 1995: A diploma in Textile Engineering was introduced. District Textile Institutes were renamed as Textile Institutes.
- In 1998: Primary Textile and RMG education started in private universities.
- 2005 to 2006: Two public universities (DUET and MBSTU) started offering B.Sc. in Textile Engineering.
- 2007 to 2011: 5 Textile Institutes turned into Textile engineering college.
- In 2011: Textile College became Bangladesh University of Textiles (BUTex). Modernization of textile education began. 5 other weaving schools also developed and started offering B.Sc. degrees in textile engineering education. BUTex and MBSTU as the first universities to offer M.Sc. in textile engineering.
- In 2012: the National Institute of Textile Training Research and Design (NITTARD) also started offering textile engineering courses. Also, the BGMEA Institute of Fashion and Technology was upgraded into BGMEA University of Fashion and Technology (BUFT), as the nation's first solely textile-based private university.
- Till 2015: The Ministry of Textile and Jute established seven textile engineering colleges; about 480 students are trained from each college.
- Till 2018: The Ministry of Textiles continued a total of 12 projects for textile education.
- In 2019: The "Feni Textile Engineering Institute" was accepted as another proposed project.
- Till 2020: A total of 10 Projects are being Under Scrutinizing for future development.

3 Employment in Textile and Apparel Industry

In Bangladesh, over 80% of foreign exchange earnings come from the textile sector, so this source is the backbone of the country's economy. In 2017, the growth rate of the garment industry was 7.89%, later it increased to 11.49% in 2019. However, due to Covid-19, the growth has been declining since then (Table 1).

Table 1 Number of Students available in each level, and the number required to cover the difference

Education level	Ph.D.	M.Sc.	B.Sc.	Diploma	Total
Existing	25	188	5034	5440	10,687
Required	1087	4708	19,996	44,863	70,654
Gap	1062	4520	14,962	39,423	59,967

At the beginning of garment industrialization in Bangladesh, inadequate experts were coming from educational institutes. The BUTex had a capacity of training only 40 graduates a year at early times. A total of 6203 graduates completed formal textile education in Bangladesh by 1999–2000. 883 were B.Sc., and 1674 were textile graduates (studied from previous DOT institutes and textile colleges). Although, within the next 7–8 years, the scenario got better. Some unofficial research tells that till the year 2006, a total of 16,216 graduates has completed their textile-related studies, which starts from Vocational level to University Graduate level education. The institutions' annual combined intake of students is nearly 4710, and the above institutes have a total enrollment of around 10,500. Depending on individual institutes, the graduation rate of enrolled students is relatively high (71–90%), and the dropout rate is very low (2–4%).

4 Existing Skill Gap in Bangladesh Textile Industry

One teacher is recommended for every 12 engineering students, but this student-teacher ratio is difficult to meet as there are inadequate teachers at service. That is why one person has to conduct multiple classes as a guest teacher. Some diploma institutes are also upgraded into Graduate Colleges, where teachers from diplomas are previously promoted to take bachelor or master's classes. Although new teachers are being hired occasionally, they deprive of facilities and support in their work. Despite their best intention and efforts, these recruits cannot give the students their best without enough training (Table 2).

According to the Annual Report 2018–2019 of the ministry of Textiles and jute, there are seven textile engineering colleges, from each college, about 480 students are trained. Also, under the Department of Textiles in Bangladesh, there are 40 vocational and three textile institutes. In addition, the textile engineering department is available in 21 private universities in Bangladesh. This large educational system produces about 188 textile postgraduates, 4580 textile undergraduates, and 2616 SSC vocational trained employees every year.

There is insufficient in-service training for employees or technical freshman people at most educational or training institutes that offer skill development for the textile sector. However, they have various attractive courses for students to become textile engineers, i.e., B.Sc. in textile engineering and diploma in textile engineering. Some of the fields' majors are preparing cotton fibers, spinning/yarn manufacturing,

Table 2 Different Occupational Level and Categorization of Employee Skills at Industries [2]

Occupation categories	% Of skilled employee	% Of semi-skilled employee	% Of unskilled employee
Managerial	96	4.17	0.17
Administrative	75	20.45	4.48
Supervisor/line leader	89	10.53	0.00
Engineer (mechanical, electrical, textile, other professional)	72	26.05	2.39
Machine operator	74	26.38	0.00
Production operator	73	18.04	8.77
Dyeing operator	66	33.76	0.00
Systems and maintenance operator	68	32.50	0.00
Purchase	71	26.37	2.20
Textile AutoCAD	71	19.17	0.00
Quality controller	79	20.55	0.17
Finishing	72	28.15	0.00
Packing	69	30.58	0.00
Knitting operator	62	38.00	0.00
Welder/electrical	55	45.05	0.00
Dryer	66	33.94	0.00
Mixer man	69	31.25	0.00
Lab boy	88	11.54	0.00
Fitter	67	33.33	0.00
Weaving operator	57	43.15	0.00
Lab technical	73	23.44	3.13
Spinning operator	65	34.78	0.00
Others	75	23.80	1.20
Total	73	20.96	5.96

weaving/knitting, dyeing, and finishing the textiles. The minimum qualification of trainees to enter each educational level can be SSC for Diploma, HSC for B.Sc., and M.Sc. for other training courses. The internship could be arranged at the end of the course for at least 15–20 days.

5 Textile Education Level in Bangladesh

There are three categories of Textile Education in Bangladesh—Vocational, undergraduate, and postgraduate. Besides, different institutes offer various courses, diplomas, and professional training available for textile-related subjects [3].

5.1 Vocational Level

Five hundred secondary schools, 64 vocational training institutes, and 12 technical training centers offer SSC-level vocational courses in different sectors. In addition, we have 40 textile vocational institutes in 40 different districts under the Department of Textile Bangladesh. The Bangladesh Technical Education Board (BTEB) implements the curriculum, and the Department of Textiles provides the Entry-level Textile Education. In addition, there is Class 8 equivalent vocational training also provided by 11 NGOs and private-run centers. Also, to offer textile and RMG courses, the Ministry of Labor and Employment (MOLandE) operates five technical training centers under the Bureau of Manpower, Training, and Employment (BMET).

5.2 Undergraduate Level

Many institutes, specialized universities, and colleges provide higher education in textile and courses on spinning, weaving, knitting, dyeing, printing, finishing, and apparel or garments, respectively (Table 3).

Under the Ministry of Textiles, Govt. of Bangladesh, the textile department also controls six textile engineering colleges (providing four years B.Sc. in textile engineering) and six textile institutes (offering four-year diploma courses).

Mid-level Textile Education in Bangladesh also consists of Diploma in Textile Engineering and manufacturing technology. The Diploma in Textile Engineering Education program is provided by the Department of Textiles (DoT). Currently, DoT provides a Diploma in Textile Engineering at only 3 Textile Institutes, Textile Institute in Tangail, Dinajpur, and Barisal. Also, four more institutes are under construction by the government. Between 1996 and 2010 a total number of 5908 Diploma engineers passed from government institutes. In addition, under Bangladesh Technical Education Board, 68 Private Textile Diploma Engineering Institutions have a seating capacity of 5440 every year.

Table 3 List of public and private institutions providing courses on disciplines related to textile engineering

Public Institutes	Bangladesh Textile University (previously known as CTT)
	Begumgonj Textile Engineering College
	Pabna Textile Engineering College
	Chittagong Textile Engineering College
	Bangladesh Institute of Textile Technology
	Abdur Rab Serniabat Textile Engineering College
	Mawlana Bhashani Institute of Science and Technology
	Dhaka University of Engineering Technology (DUET)
Private Institutes	Ahsanullah University of Science and Technology
	Daffodil International University
	Prime Asia University
	Southeast University
	BGMEA University of Fashion and Technology (BUFT)
	Atish Dipankor University
	World University of Bangladesh (WUB)
	Green University Bangladesh
	City University
	University of South Asia (UniSA)
	Bangladesh Textile Engineering College
	Shamoli Textile Engineering College
	Institute of Textile Training Research and Design (NITTRAD)
	People's University of Bangladesh
	Bangladesh University of Business and Technology (BUBT)

5.3 Postgraduate Level

Different public and private universities impart postgraduate degrees in textile engineering or related disciplines. As per the University Grant Commission (UGC) data, 12 Private Universities have approved Postgraduate Textile Engineering/Technology Education. Prime Asia University offers MBA in Textile and Apparel Marketing, BGMEA University of Fashion and Technology (BUFT) provides MBA in Apparel Merchandising and a master's program in fashion designing and textiles. Daffodil University offers M.Sc. in textile engineering; the Bangladesh University of Textile has also started M.Sc. in Textile Engineering, and they focus on four specializations like yarn manufacturing, fabric manufacturing, wet processing, and garment manufacturing.

5.4 *Professional Training Institutions*

In this area, various professional training is offered by some institutes. Such as the Institute of Textile Training Research and Design (NITTARD). It has become a public-private partnership education and training institute by going through a massive restructuring. Moreover, Bangladesh Textile Mills Associations (BTMA) is running it now. Its five departments are yarn manufacturing, fabric manufacturing, wet processing, apparel manufacturing, and quality control. With sufficient training types for practical training, it claims to offer good quality training and textile education. Previously, it offered only one year of training courses, but from 2011 it started offering textile engineering courses. It has academic collaboration with the University of Bolton and Niederrhein University of Germany. The International Textile Institute of UK accredits this One-Year certificate course of Textile Institute (LTI) Standard.

BTT Institute of Technology offers professional training and conducts several short courses, including knit dyeing and finishing, fabric technology, textile management, merchandising, garment washing technology, quality control, and lab technology. In addition, they have faculties from renowned organizations, i.e., MandS, Opex, Beximco, and SGS.

One more institution adding to this list is Bangladesh Sericulture Research and Training Institute (BSRTI), under the Ministry of Textile and Jute. According to available data, it provides Post Graduate Diploma in Sericulture (PGDS) and Diploma in Silk Technology and offers diplomas for undergraduates. It also provides training on sericulture and research mulberry seed and silkworm's species development.

6 Expansion of the Current Textile Education Compared to the Current Market Demand

The textile industry has seen enormous growth in the last few years, and huge investments were made in this sector; it is said that the investment in the primary textile sector is worth 5 billion USD. The spinning sector can now fulfill almost 100% domestic yarn demand and 85–90% yarn demand for export-oriented knitwear. Keeping pace with the expansion of the textile industry, the number of institutions providing textile education has increased widely over the years, and some considerable textile education providing institutes has gone through significant restructuring.

In 1973–74 there were no B.Sc., M.Sc. or other courses available, and in Textile College, there were only 556 students who came out taking the diploma course. Nevertheless, each year, almost 188 postgraduates and 4580 undergraduates in the textile are produced from the different public, private institutes. Also, around 70,000 professionals work in positions like managers, merchandisers, engineers, and others in the current industrial pipeline. However, the expansion is still not up to the mark compared to industry growth and industry demand.

7 Issues Regarding Existing Textile Education

Some issues regarding current textile education in Bangladesh hinder the development of qualified human capital for the industry (Fig. 2).

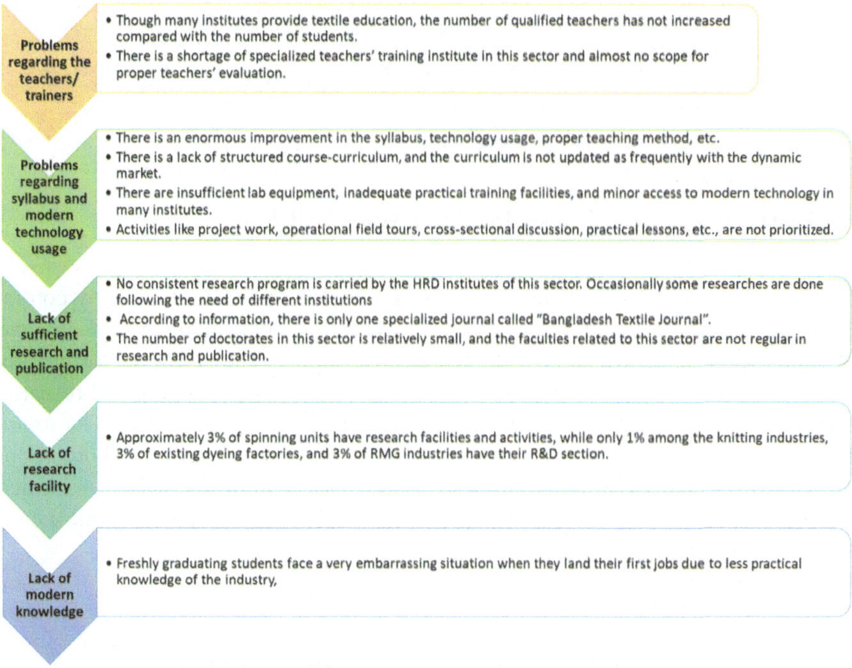

Fig. 2 Issues regarding existing textile education

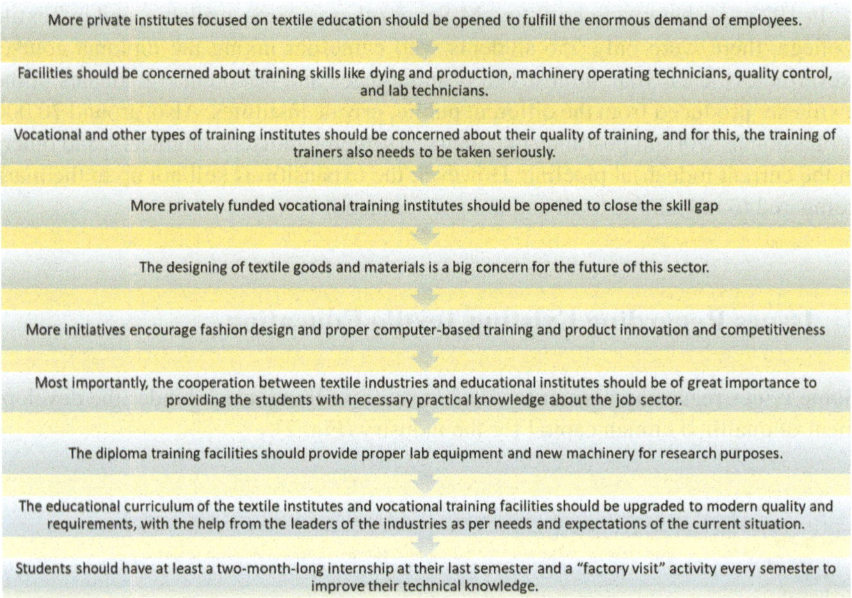

Fig. 3 Initiatives to be taken to close the skill gap

8 Initiatives to be Taken to Close the Skill Gap

Bangladesh is entirely dependent on importing textile materials, especially cotton. Also, productivity depends on technological improvement, manpower skills, and essential resource facilities, i.e., gas and electric supply. The textile sector already has a substantial existing skill gap. Although some advanced training and research institutes are increasing the workforce for the sector, that is not enough. The future of this sector is auspicious, but it needs qualified people and closing the shortages to ensure growth. The existing facilities are inadequate to fulfill the needs for skilled manpower. Therefore, to meet the requirements of skilled professionals in this sector, the following initiatives should be taken [2] (Fig. 3).

9 Present and Future Projects for Textile Education of Bangladesh

There are more than 28 public and private textile educational institutes in Bangladesh each year, and these institutes provide several almost ten thousand graduates (Table 4).

Table 4 Current and proposed projects under the Ministry of Textiles [4]

Project name	Estimated allotment according to DPP
A. Continuous Projects	
Shaheed Abdur Rab Senabiyat Textile Engineering College	12,200.00
Begumgonj Textile Engineering College	4334.82
4 new Textile Institutes	28,473.00
Jhinaidah Textile Engineering College	9921.00
Gournadi Textile Institute	6213.50
Gopalganj Sheikh Rehana Textile Engineering College	12,973.17
Vola Textile Engineering Institute	6499.00
Jamalpur Sheikh Rasel Textile Institute	6250.00
Rangpur M.A Wazed Miah Textile Engineering College	10,748.24
Jamalpur Sheikh Hasina Textile Engineering College	11,142.59
Manda Shaheed Kamruzzaman Textile Institute	7462.49
Sirajganj Begum Amina Mansoor Textile Engineering Institute	8746.32
Total	124,964.13
B. Accepted Project (at 2019)	
Feni Textile Engineering Institute	9782.80
C. Projects Under Scrutinizing (till 2020)	
Sylhet Textile Engineering College	11,425.99
Sylhet Textile Engineering Institute	10,313.54
Sunamgonj Textile Institute	10,052.67
Faridpur Textile Institute	10,953.42
1 Vocational Institute	19,298.32
Lalmonirhat Textile Institute	9412.88
Sharasitma Textile Institute	11,068.51
Kapashia Textile Engineering College	11,695.00
Netrokona Textile Institute	9803.88
Jessore Textile Institute	9803.88
Total	113,828.1
A + B + C	148,575.03

From the Ministry of Textile's yearly budget, we can establish and upgrade 18 more textile colleges and institutes. The present government has allocated a sum of 148,575.03 Crore BDT in previous academic years.

10 Laboratory Set-Up Problems and a Proposed Solution

Laboratory machines are required in textile colleges and institutes to develop lab facilities. For that most expensive and high-volume scale industrial machines are being purchased for small laboratory purposes. However, after that, the educational institutes find it difficult to operate and maintain these high-functioning machines. For example, the standard power supply required for these machines is about 380 V or around 7 kW and needs to be under maintenance by skilled technicians. Also, rare and expensive raw materials are needed to keep these machines in running condition. Most importantly, these industrial machines produce a high pitch noise (almost 100 dB), damage the ears, and are unsuitable for an educational environment. Within a few years of purchase, these machines turn out to be just some dummy samples to show the students, but they cannot watch them in running conditions. However, with the overall cost of 2–3 of these machines, any institute can cover up the complete lab equipment set if they purchase lab-scale machines [4] (Fig. 4).

As we can see from above, a considerable workforce, energy, and raw materials are required to operate the expensive industrial-scale machine. However, it is cheap for the lab machine, can run by the institute's internal power supply, and the lab assistants themselves can efficiently operate it. However, as far as we know, textile institutes still prefer these high production-oriented industrial machine lines. Therefore, it is miserable that the students who are supposed to learn cannot do a single test using them, although the universities spend millions of dollars purchasing and setting up a lab environment. As a result, most undergrad students fail to develop quality project work; some motivated students try their best but face difficulty researching independently. Because without the opportunity of university laboratories, they have to find some industrial factory or commercial textile laboratories and the personnel there is very tough to convince. Even if some students manage to get a chance to access it, they have very limited project time left, and sometimes it is not enough to finish the research. So eventually, they have to submit incomplete, fabricated data. Moreover, sometimes the factory location is quite far from the university area, so it is

Fig. 4 Some images of spinning and weaving lab of Department of Fabric Engineering, Bangladesh University of Textiles, Dhaka, Bangladesh. Images courtesy of Shamima Akter Smriti

also challenging to travel, takes a long time on the road, and spends lots of money on transportation costs. These are the significant reasons why bachelor students cannot bring a fruitful thesis work, and the faculty members also already know all these facts, so they got no other choice but to overlook the matter.

11 Scopes for Bangladeshi Students on Higher Education or Research Opportunities

Bangladeshi students have the opportunity at many different institutions around the world that offers various courses on Textile and Apparel, but as it is a highly diversified subject and varieties of applied degrees are offered, including major in Textile Engineering, Apparel Manufacturing, Knitwear Manufacturing, Fashion Technology, Textile Management, and Apparel Merchandising. So, before traveling from Bangladesh, a student has to decide Where to study and What to study.

In the case of Textile Studies offered for Bangladeshi students, we can categorize all the universities into three major blocks [5]:

1. US/Canada
2. UK/EU (Europe)
3. Asia/Australia

That is because, among these three regions, their academic curriculum, education quality, offered scholarships, and admission requirements are different from each other.

11.1 USA

The USA or Canada is a top choice for students from Bangladesh who want to pursue their higher education in Textile Studies, as because the North American certificate provides a great value on the job market, it is a great place to stay. However, compared with other engineering fields, universities in the US offering above mentioned textile courses are still relatively low. Same situation in Canada, where mainly three or four universities have direct majors in textile. Here is a list of American universities that are most popular among Bangladeshi students to study their higher education on Textile:

1. North Carolina State University (NCSU)—Considered among the top 100 universities of the world, the College of Textile of NCSU is one of the best textile institutes. Admission at NSCU as a graduate is very competitive, and one needs to have at least a CGPA of 3.0 out of 4 in bachelor's and a minimum 65% GRE Score to be eligible to apply. Also, a TOEFL score of 80 with a minimum of 18 in each group of subjects must sit on the online test.

2. Kansas State University (KSU)—The admission qualification to KSU is also quite difficult and requires a pretty good academic profile to convince a supervisor.
3. University of Philadelphia (PhilaU)—The textile education wing under the University of Philadelphia is called "Kanbar College of Design, Engineering, and Commerce". For admission to this excellent academic college, a student needs to have academic transcripts, two recommendations letters, official GRE scores, a Statement of Purpose, and impressive academic background.
4. University of Missouri, Columbia (UM)—Admission to this excellent university also needs a high GRE and English proficiency certificates (TOEFL).
5. University of Nebraska-Lincoln (UNL)—Bangladeshi Students must have at least a CGPA of 3.0 out of 4 to apply to this university. Also, the minimum Graduate Record Exam (GRE) scores required for textile science are at least 143 in Verbal, 155 in Quantitative and 200 in Analytical Writing.
6. Florida State University—The "Department of Retail, Merchandising and Product Development (RMPD)" offers nearly 800 majors, one of the largest programs in the United States. It provides an industry-oriented curriculum, and the graduates are immediately hired and can contribute to retail and retail-related companies. Florida Retailers are providing one out of every five jobs in the US. In addition, the RMPD has a connection with over 200 companies, government agencies, and trade organizations that offers RMPD graduates various internships and job placements. So, an RMPD major from FSU can be considered an excellent asset for Bangladeshi in the American retail industry.

Apart from the names above, more than 20 universities, colleges, and institutes provide excellent textile, apparel, and fashion courses. However, it is challenging to get some funding from them. So, if someone can bear their educational costs, these places are worth trying.

11.2 Canada

Canada is less popular than the USA among students pursuing higher studies in textile and apparel. Nevertheless, these two universities below can be considered as a great choice:

1. University of Manitoba- The Department of Textile Science offers an M.Sc. degree program for graduates. Students here can research textile materials and product development, Consumer behavior, Textile product marketing, Supply chain management, Biomedical textiles, and Nanomaterials. In addition, the department has many laboratories, including CAD, Textile testing, and medical textile labs, which offer great scopes for work experience.
2. University of Alberta- The department of human ecology also offers a prestigious M.Sc. degree in Textiles and Clothing.

Admission procedures in Canada are a bit different from the USA, as institutes prioritize B.Sc. results here. So, a minimum CGPA of 3.75 out of 4 is required to get a scholarship or financial support.

11.3 UK

To talk about Textile education in the European Union, primarily the UK and Germany, come into our concern. Some UK-based universities offer the prestigious "Commonwealth Scholarship" for Bangladeshi students and another special scholarship named "Equity-Merit Scholarship" for Bangladesh, Tanzania, and Uganda. The Education Ministry of Bangladesh offers the Commonwealth Scholarship and Fellowship Program (CSFP) every year. These are only given to students with exceptional research works and academic records, and they are entirely merit-based awards. University Grants Commission of Bangladesh (UGC) publishes yearly updates about CSFP on their website:

1. University of Manchester—Manchester is called the City of Science in Europe, and it is one of the ancient universities of the world. It is also ranked among the top 30 universities in the world. Manchester University offers the above-mentioned Commonwealth Scholarship, and already, several Bangladeshi students have been enrolled here with the help of this scholarship. Nevertheless, to match the eligibility criteria, a minimum IELTS score of overall 6.5 or 7.0 in some courses is required from applicants. In addition, having publications in international journals may help to get better feedback.
2. University of Bolton—This university also offers various postgraduate degrees in textile, like a master's course emphasized on Technical Textiles in Advanced Materials. Also, at the "Institute of Materials Research and Innovation (IMRI)", several research projects are running under the direct supervision of renowned professor Dr. Shuvash Anand. Compared with UK standards, tuition fees at the University of Bolton are one of the lowest. Although they do not provide privately funded scholarships to Bangladeshi students, they offer waivers according to entry qualifications and yearly progress. Besides that, outside scholarships like the Commonwealth Scholarship can be approached by applicants to cover their total tuition fees along with living allowances. Also, Bangladeshi students can study here with the help of the UNIDO Fellowship Program.
3. University of Leeds—This university offers master's programs in six majors: Design, Advertising, Fashion Enterprise and Society, Textile Innovation and Branding, Advanced Textiles, and Medical Textiles. Also, Ph.D. is available on Medical and Advanced Textiles. In addition, for Bangladeshi students, the university offers a postgraduate research scholarship. Nevertheless, applicants must have either a first-class degree or an excellent academic performance record as the top 20% of their group to be eligible for that.

4. Manchester Metropolitan University—This university can be another option in Manchester city for higher education in Textiles. Some scholarships and Waivers are being provided from MMU, but an applicant needs to look for scholarships offered through the British Council to get a fully funded stipend.
5. University of Westminster—At Westminster, the most promising course to look out for is the MA in Fashion Business Management. When it comes to supporting Bangladeshi students, this university is one of the most generous scholarship providers in the UK. Every year many students from Bangladesh apply for the Westminster International Scholarship. WIS pays for the full tuition fee, accommodation, and living expenses and gives funding for the airfare from Dhaka to London.

11.4 Germany

German Universities do not charge tuition fees, so getting a scholarship is difficult since the students must manage their stay and accommodation costs and cover health insurance. To get a full funding studentship in Germany, the "German Academic Exchange Service (DAAD)" is the only source of help. From the DAAD website, the requirements of the scholarships and application procedures can be easily collected. A good number of students from Bangladesh are being sponsored by DAAD every year:

1. Niederrhein University—Considered as one of Europe's most significant study locations for textile education, the faculty of textile and clothing technology holds the most significant number of international students in whole Europe. Niederrhein University is a friendly place for international students as around 25% of students here are from 25 different countries. Although, it is not included in the DAAD scheme of universities at present. The textile and clothing sectors are the features at Niederrhein with a variable range of degrees to offer under some of the best professors.
2. The Technical University of Dresden is another famous university in Germany for textiles, also financed by DAAD.

Apart from the names mentioned above, the "University of Ghent" in Belgium and the "Boras University" in Sweden are also famous in Europe for their postgraduate textile studies.

11.5 China and Japan

Recently, many scholarships are being offered to Bangladesh from famous Asian Universities, but an applicant must have a good research background and a rich student profile to be eligible. One of these regional universities' essential advantages

Fig. 5 A graduate research student at the College of Textile Science and Engineering, International Institute of Silk, Zhejiang Sci-Tech University, Hangzhou, China, who completed his education from BGMEA University of Fashion & Technology, Bangladesh

is that the students do not need to sit for GRE tests, which many consider troublesome. At present, textile education is becoming essential in China, as they are offering a high number of scholarships in these research fields for higher studies. To receive excellent standards in textile education, a student can find many qualified universities and institutes in China (Fig. 5).

Shinshu University offers advanced postgraduate courses on textile engineering and material science. The Department of Textile Science and Engineering has two divisions, the Division of Creative Engineering and the Division of Chemistry and Materials. Moreover, many academic scholars from Bangladesh are being sponsored every year by offering the famous "Monbukagakusho Scholarship" in Japan.

11.6 Korea

As most universities provide enormous funds for theoretical and applied research purposes, South Korea has become a popular place for Bangladeshi students in recent times. Therefore, an applicant has to match research interest and find a suitable supervisor because, for research students, South Korean universities have regular funding offerings:

1. Yeungnam University is probably the only university in South Korea that provides course offerings in Textile, clothing and fashion Engineering and materials engineering. So, YU can be considered as one of the few comprehensive textile universities in Korea.
2. Seoul National University—In South Korea, SNU is considered one of the best research universities. Decorated with an advanced course curriculum, the Clothing and Textiles department includes basic textiles, fiber properties, clothing history, performance and ergonomics, design, and trading.
3. Kyung Hee University—This university offers a nationally ranked program through Clothing and Textiles. This program teaches inter-reactions between human beings and the clothing environment and seeks to produce capable, creative scholars and professionals.

11.7 Australia

Some Australian universities offer lucrative scholarship programs to Bangladesh, and they can be one of the numerous choices, as they are also conducting various research projects on advanced textiles. However, these universities mainly focus on core textile fields, i.e., advanced fiber, nanotech, and material sciences. Therefore, to obtain a postgraduate research place, a student must have an attractive research background with significant impacts to qualify for a scholarship. Moreover, for international students every year, the "International Postgraduate Research Scholarship (IPRS)" is offered that sponsors total tuition fees, residence in Australia, establishment costs, health insurance, and even expenses for spouses and children. Targeting research scholars for postgraduate studies from Bangladesh, IPRS is considered the most prestigious Australian scholarship.

1. RMIT University—When textile education is concerned, it is one of the best universities in Australia. Students can either choose their degree by research or by the taught program in this renowned research university. A minimum IELTS band of overall 6.5 with no individual band below 6.0 and an excellent academic record is needed to secure a place in the school. The RMIT School of Fashion and Textile is decorated with around 20 high-profile faculties with various research areas.

Apart from the above names mentioned, the "University of Technology" in Sydney, the "Australian National University", "Deakin University" and "University of South Australia" can be some other places to be considered as a Bangladeshi student or researcher.

12 The Scenario of Co-operation Between Bangladesh and International Universities

The academic strategy of educational institutes needs a global approach. Sharing the same academic direction, many leading international universities and organizations have strategic agreements and partnerships with universities in Bangladesh. The global mainstreams of textile, apparel, fashion, and business to integrate their students. Through exchange programs, international collaborations allow students to experience the "study abroad" options. By getting selected for the exchange program, this initiative provides the students with an excellent opportunity. It encourages them to broaden their vision and understand different cultures by interacting with students from various geographies. The university's international linkage allows the students to participate in international competitions, seminars, research, exhibitions, and other events to provide academic progress. Overseas universities with strategic agreements and partnerships with Bangladeshi universities can pursue higher education with scholarships available by many students. Bangladesh also attracts international students to come and experience its academic and cultural richness. Students from overseas institutes and universities understand the Bangladeshi market and its dynamics and develop a valuable insight into Bangladeshi culture, arts, and crafts through the exchange programs. Here are two examples of co-operation between two universities in China and Bangladesh [6, 7] (Fig. 6).

Fig. 6 MoU signed between BUFT and Zhejiang Sci-Tech University, China

Fig. 7 MoU signed between BUFT and Wuhan Textile University, China. *Source* Edu Icon (August 08, 2015)

Prof. Dr. Engr. Ayub Nabi Khan, CText FTI, Pro-Vice-Chancellor of BGMEA University of Fashion and Technology, visited Zhejiang Sci-Tech University, China, in April 2017. During this time, there was a fruitful discussion on the status of the present Education Collaboration and the new areas of possible research and educational partnership with ZSTU and BUFT. In his visit, the primary purpose was to promote, consolidate and strengthen collaboration in Apparel, Fashion, and Textile education for Bangladesh's apparel industry. During the visit, a new Memorandum of Understanding (MoU) was signed between ZSTU and BUFT. This provided an excellent opportunity for BUFT graduates to do their master's and Ph.D. at Zhejiang Sci-Tech University, China (Fig. 7).

The Wuhan Textile University of China signed a Memorandum of Understanding (MoU) with BGMEA University of Fashion and Technology (BUFT) in April 2015 at the WTU President's Office. The two institutions aimed to promote and strengthen collaboration in Textile, Apparel, and Fashion education. Prof Wei Yiliang, President of Wuhan Textile University, and MP Tipu Munshi, former BGMEA president and BUFT Board of Trustees, signed the agreement on behalf of their respective universities. In addition, both the universities agreed on exchanging students, researchers and faculty, transfer credit and providing scholarships to the students. They also planned to hold joint exhibitions and fashion design competitions and develop and design curriculum.

13 SWOT Analysis of Textile Education in Bangladesh

SWOT analysis has been done as a strategy, and the following measures can be taken to improve the quality of Textile Education in Bangladesh (Fig. 8).

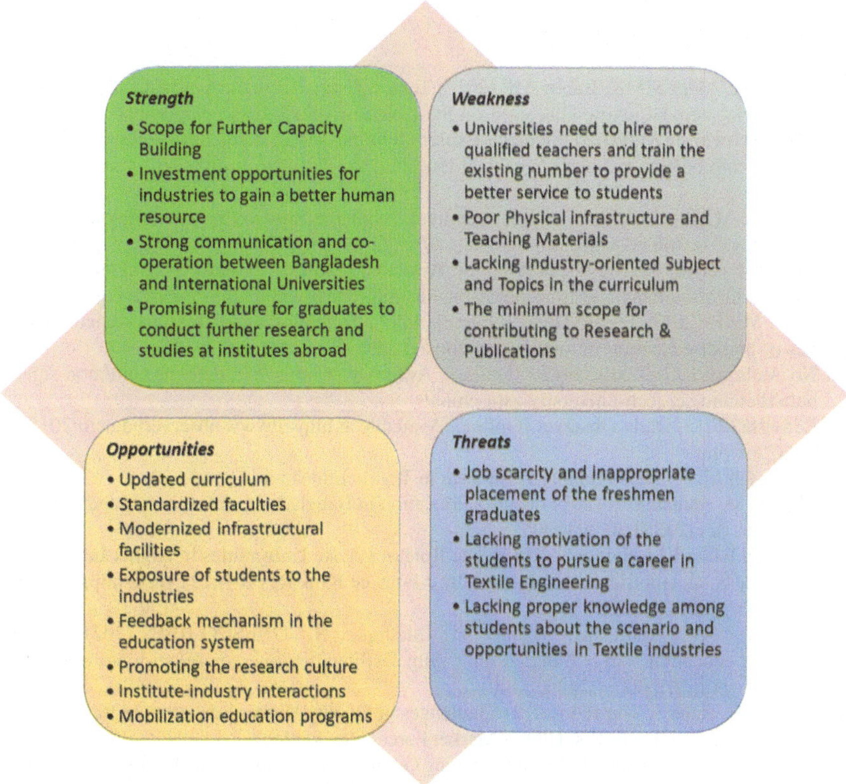

Fig. 8 SWOT Analysis of Textile Education in Bangladesh

14 Conclusion

In the present world, almost every natural resource can be obtained in a limited time by exchanging money, except only the "Human Resource," which is increasing demand day by day and may not be readily available so quickly at the time of need. So, any business or industry can face difficulties or lose competitiveness due to a lack of skilled human resources. The Scenario of Textile education in Bangladesh has grown considerably in quality and quantity over the years, although there is still a large demand gap for textile-educated people in the industry [8–23]. To resolve this massive existing and upcoming crisis, assuring and improving quality in educational fields is the topmost priority. With proper commitment and effective initiatives, which will bear significant impacts on the future of the Textile scenario in Bangladesh.

References

1. Iqbal MA, Akter MMK, Rabby MF (2017) Textile Today [Online]. Available at https://www.textiletoday.com.bd/textile-education-on-the-table-at-its-100-years-completion/
2. Ahmed N, Parvin M (2018) Financial Express [Online]. Available at https://www.thefinancialexpress.com.bd/views/analysis/a-critical-look-into-the-skill-gap-in-the-textile-industry-1544370149
3. Siddique MAR (2017) Textile Today [Online]. Available at https://www.textiletoday.com.bd/structure-of-textile-education-in-bangladesh/
4. Ahmed T (2018) Textile Focus [Online]. Available at https://textilefocus.com/present-trend-textile-education-bangladesh-proposal-establishing-combined-textile-research-center/
5. Akter MMK (2017) Textile Today [Online]. Available at https://www.textiletoday.com.bd/scopes-of-higher-education-on-textile-and-apparel-2/
6. No Author (2017) Textile Focus [Online]. Available at https://textilefocus.com/mou-signed-buft-zhejiang-sci-tech-university-zstu-china/
7. Eduvista (2015) Daily Observer [Online]. Available at https://www.observerbd.com/2015/04/07/82268.php
8. Das S, Ghosh A (2005) Quality Assurance in Textile Education
9. Monem M, Baniamin H (2011) Higher Education in Bangladesh: status, issues and prospects. Pak J Soc Sci (PJSS) 30:293–305
10. Anwarul Islam KM, Salma U (2016) The Role of Private Universities in Higher Education of Bangladesh: an empirical investigation. Int J Finance Bank Res 2(4):121–128. https://doi.org/10.11648/j.ijfbr.20160204.11
11. Varghese NV, Alam M, Haque M, Siddique S, UNESCO-IIEP (2021) Private higher education in Bangladesh. http://lst-iiep.iiep-unesco.org/cgi-bin/wwwi32.exe/[in=epidoc1.in]/?t2000=024260/(100)
12. Sarkar SH, Rana S, Zitu RA (2012) Challenges of Quality Higher Education in Bangladesh: a study on public universities. Chives Market Forces Res J 8(4)
13. Mohsin M, Aktar Kamal M (2012) Managing Quality Higher Education in Bangladesh: lessons from the Singaporean and Malaysian Strategies and Reforms. Int J Biomet 7:59
14. Andaleeb S (2003) Revitalizing Higher Education in Bangladesh: insights from Alumni and Policy Prescriptions. High Educ Policy 16:487–504. https://doi.org/10.1057/palgrave.hep.8300036
15. Rabbani G, Chowdhury S (2014) Quality of Higher Education in Bangladesh: governance framework and quality issues. Beykent Üniversitesi Sosyal Bilimler Dergisi 7
16. Ahmed JU (2016) Massification to Marketization of Higher Education: Private University Education in Bangladesh. Higher Educ Future 3(1):76–92. https://doi.org/10.1177/2347631115610222
17. Asadullah MN (2006) Returns to Education in Bangladesh. Educ Econ 14(4):453–468. https://doi.org/10.1080/09645290600854144
18. Sarkar SH, Hossain SZ (2018) Higher Education Systems and Institutions, Bangladesh. In: Teixeira P, Shin J (eds) Encyclopedia of International Higher Education Systems and Institutions. Springer, Dordrecht. https://doi.org/10.1007/978-94-017-9553-1_499-1
19. Dolby N, Rahman A (2008) Research in International Education. Rev Educ Res 78(3):676–726. Accessed June 27, 2021. http://www.jstor.org/stable/40071141
20. Hébert Y, Abdi AA (2013) Critical perspectives on. Int Educ. https://doi.org/10.1007/978-94-6091-906-0
21. Nwaokaomah A (2010) Strategies for promoting entrepreneurship opportunities in Clothing and Textile Education. JHER 13:42–49
22. Power J (2007) An examination into textile education by active research. In: Proceedings of the 85th Textile Institute World Conference. The complete set of world conference proceedings. The Textile Institute, Colombo, Sri Lanka
23. Heffernan S (2015) The Value of Textile Education and Industry Partnerships. J Text Des Res Pract 3(1–2):65–85. https://doi.org/10.1080/20511787.2015.1184407

Mohammad Fahim Uddin was born in Dhaka, Bangladesh. He is currently pursuing his post-graduation from Zhejiang Sci-Tech University, one of the best institutes in China that specializes in textile studies. With a major in Material Science and Engineering, his field of research is Nanotechnology and Biomedical Textiles. He completed his graduation in 2018 from the well-reputed BGMEA University of Fashion and Technology (BUFT) in Bangladesh, majoring in Apparel Manufacturing and Technology. After that, he had work experience as a Merchandising Executive with Multinational Brands like Walmart, PVH, and Amazon. Apart from institutional and professional achievements, he also completed various training from the Finance Ministry of Bangladesh and participated in Inter-university Debate Competitions. Currently, he is living in Hangzhou, the beautiful capital of Zhejiang Province of China.

Textile Education in Pakistan

Gohar Ali Hayat, Muzammal Hussain, Muhammad Qamar Khan, and Zafar Javed

Abstract This article investigates to outline the education and business trends of the Textile Sector in Pakistan. This article's fundamental purpose is to highlight the importance of educational institutes' contributions in textile in Pakistan, which helps the Textile sector endure a competitive edge worldwide. Moreover, this article highlights those domains and how Pakistan should excel to enrich textile education according to worldwide best benchmarks. This article can be beneficial for our youth as well, who are interested in the Textile sector. They can get good insights about prestigious and reputable Pakistani textile institutes to attain excellent quality education. A complete course outline and module have been discussed in textile education like spinning, weaving, textile chemistry, fashion, textile design, knitting, and garment manufacturing. This article also demonstrates the critical rules of outcome-based education in which educationists came to know about its importance and encourage the students to get in this field which will raise their future. They have better opportunities to serve the world by promoting the textile industry and textile education department through innovation and research in the textile field. This article covers the valuable information related to scholarships and other support systems available to textile students and faculty.

Keywords Textile education · Textile business trend · Outcome-based education · Scholarship opportunities · Guidance of textile course modules

---

The original version of this chapter was revised: The author name "Muzammal Hussain" has been updated. The correction to this chapter is available at
https://doi.org/10.1007/978-981-16-8854-6_10

G. Ali Hayat · M. Hussain · M. Qamar Khan (✉) · Z. Javed
Department of Textile and Clothing, National Textile University Karachi Campus, Karachi, Pakistan
e-mail: drqamar@ntu.edu.pk

X. Yan et al. (eds.), *Textile and Fashion Education Internationalization*, Textile Science and Clothing Technology, https://doi.org/10.1007/978-981-16-8854-6_4

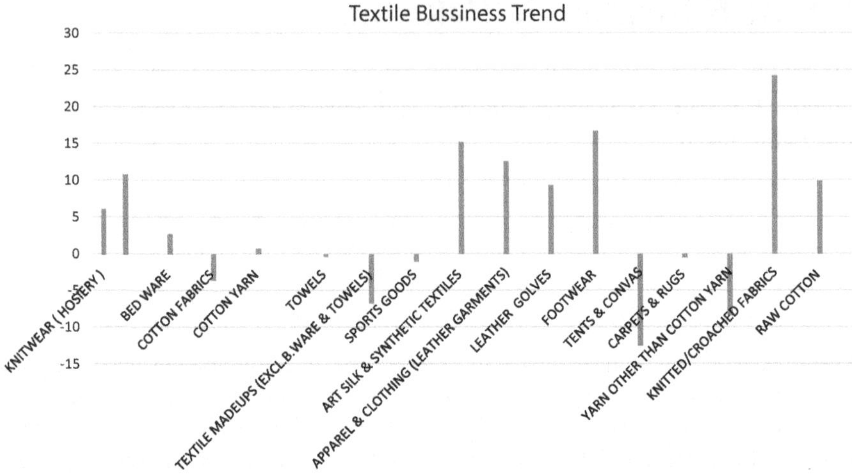

Fig. 1 Textile business trends for the year of 2019–2020 comparing compared to 2018–2019 [3]

1 Introduction

The textile industry is known as the prime manufacturing industry of Pakistan. Pakistan possesses around 13,000,000 spindles, 1,908,801 Rotors, twenty-eight thousand five hundred shuttles loom, 3,075,000 powers looms [1], and four hundred sixty-four composite units; about twenty-two units have been registered with Pakistan stock exchange under the textile business [2]. In the below-given bar chart, Pakistan's textile exports of US$ over the year 2019–2020 can be seen. However, the attributed variation in exports for each category is shown in the bar chart due to the epidemic (Figs. 1, 2 and 3).

About 45% of Pakistan's labor force is part of the textile sector in Pakistan. Moreover, the textile sectors contribute 8.5% to the GDP of Pakistan. As mentioned earlier, textile education in Pakistan has enormous importance and scope because it improves economic conditions [4]. Textile education is offered in both the public and private sectors in Pakistan; refer to Tables 1 and 2.

All Textile Universities of Pakistan are prestigious institutes that provide excellent quality textile education. Textile Universities offer various textile programs such as Textile Engineering, Textile Management, and Marketing, Textile Design to Advance textile in every textile and fashion industry aspect.

2 History of Textile Universities

NTU (the Lyallpur Institute of Textile Technology) was founded on October 12, 1959, in partnership with the cooperation of the Government of the United Kingdom,

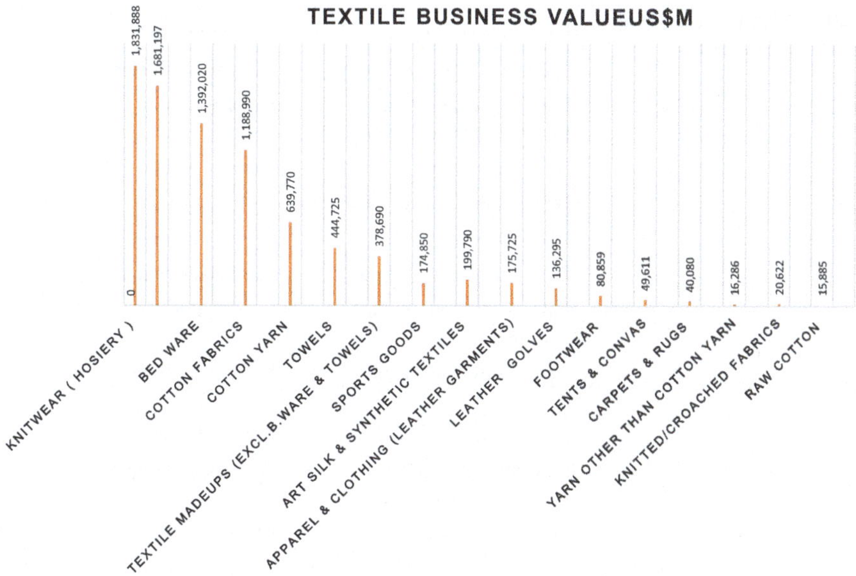

Fig. 2 Textile export in US$ million categories wise [3]

Fig. 3 Export market of Pakistan's textiles for the year 2019–2020 [3]

Table 1 Public sector textile educational institutes

University name	City of university (province)
National Textile University	Faisalabad, Punjab
Mehran University of Engineering and Technology	Jamshoro, Sindh
NED University	Karachi, Sindh
Bahauddin Zakariya University	Multan, Punjab
BUITEMS	Quetta, Baluchistan
UET Faisalabad Campus	Faisalabad, Punjab
Pakistan Institute of Fashion Design	Lahore, Punjab
Agriculture University Faisalabad	Faisalabad, Punjab
National College of Arts	Lahore, Punjab
Government College University Faisalabad	Faisalabad, Punjab
Technical Education and Vocational Training Authority Punjab	Every District in Punjab
University of Gujrat	Gujrat, Punjab
University OF Sindh	Jamshoro, Sindh
Kinnaird College for Women	Lahore, Punjab
University OF Jhang	Jhang, Punjab
Lahore College for Women University	Lahore, Punjab
University of Punjab	Lahore, Punjab
University Of Sargodha	Sargodha, Punjab
Women University Swabi	Swabi, Khyber-Pakhtunkhwa

Table 2 Private sector textile educational institutes

University name	City of university (province)
Indus University	Karachi, Sindh
Indus Valley School of Art and Architecture	Karachi, Sindh
Hajvery University	Lahore, Punjab
University of Management and Technology Lahore	Lahore, Punjab
Iqra University Karachi	Karachi, Sindh
Saylani Textile Training Center	Karachi, Sindh
University of Sialkot	Sialkot, Punjab
Textile Institute of Pakistan	Karachi, Sindh

the Federal Pakistani Government, and the textile industry. Pakistan Government gave the land of around 62 acres, the capital such as machinery and equipment were supplied by the UK government (the products mentioned earlier were not supplied free of cost). Moreover, approximately 2.5 million were contributed by Textile industries such as Lyallpur Cotton Mills, Kohinoor Industries, Dawood Foundation, Colony Textile Mills. The other government also enacted some tax laws for textile manufacturing to run the organization. The organization was associated with the University of Engineering, Lahore, in 1965, and it was established as the National College of Textile Engineering. Board of Directors was created, in which the Minister of Businesses was appointed as Chairman and the candidates of the donor companies as members to accomplish the organization's affairs. The government imposed a Cess on the Textile Industry of Pakistan to meet the recurring expenses of the university. Later, in 1965, the University of Engineering and Technology, Lahore, affiliated with the institute and renamed it "National College of Textile Engineering." In 1966, the very first batch of engineering students graduated. The institute's organizational authority was passed to the Federal Government in 1973, and the institution was renamed the "National Textile Engineering College". The National College of Textile Engineering received roughly 650 million yen from the Japanese government under the JICA program in 1992. On November 15, 2002, the college was elevated to National Textile University.

Multan has traditionally been a center of academic brilliance. Hazrat Bahauddin Zakariya (*1172–1262 AD*), a Muslim spiritual, intellectual, and saint, founded a theological institute at Multan, where intellectuals from across the world gathered to study and study. By preserving its traditional cultural legacy, Multan has always been a perfect learning hub. Consequently, in 1975, an Act of the Punjab Governmental Assembly originated the University of Multan. In 1979, the university's name was altered from University of Multan to Bahauddin Zakariya University in honor of the Great Saint.

The university has made enormous progress over the years, but some recent achievements are particularly noteworthy. Multan School of Arts was founded in 2003 to encourage art and cultural activities. The University Textile Engineering was founded in 2004 to meet the needs of the textile industry, which is a prominent industry in the area. As part of the government's policy, the university has also created three sub-campuses at Vehari, Layyah, Lodhran, and Muzaffar Gharh to deliver education to people at their doorsteps. In addition, the university has taken steps to strengthen its ties with the community and industry. As a result, the number of pupils enrolled has increased significantly.

QEC (Quality Enhancement Cell) has been formed to ensure excellent quality in Research and Academics. Numerous buildings' construction is in progress, which shall be used to render quality education to students. There is a semester system, and international conferences and seminars have become a regular academic activity. Post-doctoral research grants have been awarded to many faculty members. Recent developments include the Saraiki Area Education Centre, Institute of

Ergonomics, and Faculty of Veterinarian Sciences. The Bahauddin Zakariya Institution is Pakistan's fastest-growing public university, and it will continue to play an essential part in the country's development.

In March 1977, The NED University of Engineering and Technology was formed under the act of the Local Assembly of Sindh, after the promotion of the prior NED Government Engineering College. As a result, NED University has enlisted Pakistan's oldest institutions for instruction and creating engineering gradations. Before this, classes were held at D. J. Sindh College to address assistants for the Sindh PWD, Municipalities, and Local Boards.

To address the demands of Civil Engineers working on the project of Sukkur Barrage, College Principal C.S. Shahani made a concentrated effort to establish Engineering Degree classes on August 29, 1921. Later, an application was sent to the University of Bombay by the Commissioner of Sindh, the ex-officio president of Sindh Collegiate Association (a registered society of subscribers for education Sindh).

NED Government Engineering College was converted to NED University of Engineering and Technology was finally formed on March 1, 1977. From 50 students in 1923, the student population has grown to approximately 7000 students at undergraduate and graduate levels. The Bio-Medical Engineering faculty is located on the NED LEJ Campus, donated by (Late) Latif Ebrahim Jamal, a well-known philanthropist, for which the land and building are believed to be worth Rs. Three hundred fifty million were provided.

Mr. A. M. Akhoond was the first Vice-Chancellor of the University, and he was followed by Prof. Dr. A. T. Khan, Prof. Dr. Jameel Ahmed Khan, Prof. Dr. M. Munir Hasan, Prof. Dr. A. Q. Qazi, Engr. Abul Kalam, Prof. Dr. M. Afzal Haque, and Prof. Dr. Sarosh Hashmat Lodi, the current Vice-Chancellor.

3 Program Structure Offered to Students in Textile Education

All reputable Universities of Pakistan offer programs such as in BSc. Textile Engineering, BSc. Textile Management and Marketing, BSc. Polymer Engineering, BSc. Fashion Design and Technology, BSc. Textile Design and Technology, MSc. Intelligent and Smart Textile, MSc. Fiber Science, MSc. Advanced Materials, Ph.D. in Textile Engineering, Ph.D. in advance material, Ph.D. in chemistry, and Ph.D. in Computer science.

3.1 Course Modules for Textile Education

Courses such as textile chemistry, spinning, weaving, fashion, and textile design, knitting, and garment manufacturing are offered in Textile Engineering to pupils. In either of the above, the students will get their specialization in mathematics, English, drawing, textile engineering utilities and services, machine application are some of the subjects commonly taught by Textile Engineering students. In Engineering Design, yarn production engineering, high-performance fibers, post spinning activities, entrepreneurship/quality control systems, textile environmental and social compliances, advanced spinning techniques, woven fabric structure, and design, weaving calculations, weaving mechanism, specialty weaving, structure and design of knitted fabric, knitting preparatory process, knitting mechanism.

3.2 Course Modules for Polymer Engineering

In polymer, engineering pupils offer courses regarding polymer additives and their compounding, polymer processing, polymer recycling, polymer reaction engineering/reinforcement for composite materials, manufacturing of plastics, techniques for polymer characterization, polymer processing.

3.3 Course Module for Management/Textile Marketing and Management

In Textile Marketing and Management pupils are offered courses such as Contextual Studies, Textile Raw Materials, International Relations, Current Affairs, Organizational Behavior, and Product Costing in Textile and Apparel Industry, Textile Testing, and Quality Control, Merchandising and Operations Management, Fashion Range Planning.

3.4 Course Module for Fashion Design

The Fashion Design program consists of courses such as Sculpture, Sewing, Fashion Business, Digital Communication, Draping, Textile Basics and fibers, Pattern and Grading, Foreign Language, etc.

3.5 Course Module for Textile Design

The program of Textile Design offers courses for students such as Photography, Textile CAD, Fashion Illustration, and Development in Textile Design, Product Detail and Development, Smart Textiles, and Textile Calculations, etc.

3.6 Course Module for Advance Material

Advance Material offers Composite Technology, Technical Textiles, Nano Materials, Smart Materials, and Mechanics of Materials, etc.

4 Outcome-Based Education (OBE)

One of the most notable developments in Textile educations in recent years has been the transition to outcome-based education. This chapter also demonstrates outcomes and outcome-based education, outlines the evolution of outcome-based education, examines multiple ways in which outcomes have been interpreted, and discusses the benefits of the outcome-based approach to education.

4.1 The Introduction of Outcome-Based Education in Pakistani Universities [5]

Since being a member of the International Engineering Alliance and signing the Washington Accord, the Pakistan Engineering Council (PEC) has promoted outcome-based education.

All the international and well-reputed educational institutes support and apply the OBE system. The core purpose of this system is to enhance students' creativity instead of just making them nerds. As per this system, some goals are designed for each course at the very start, and it is mentioned clearly in course outlines which faculty members follow round the semester, so throughout and especially after the completion of education, every student should know what and why he/she has studied in any course. Moreover, it can be said that the OBE system prepares students for a practical world where they may demonstrate whatever they have learned. There was no concept of OBE back in the last few decades, but It has proved itself as an excellent approach to education and has raised our educational quality standards. So now we may say that our student has the proper knowledge and knows how to apply his/her knowledge in the practical/professional world.

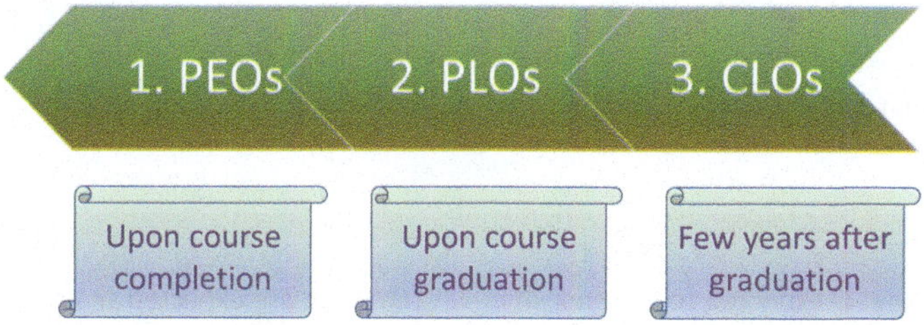

Fig. 4 Flow chart of outcome-based education [6]

4.2 *CLO and PLO*

All Pakistani Textile Universities implement Outcome-based education, and in this regard, there are specific goals designed related to the overall program and course. Therefore, PLO stands for Program Learning Outcomes; CLO stands for Course learning Outcomes (Fig. 4).

Program Educational Objectives PEO is linked to PLO. The link between each PLO and CLO is (course learning outcome). Each course is mapped to PLO to ensure the outcome of the training. The system of teaching and evaluation is often intended to achieve the outcome of education. Key Performance Indicators (KPIs) are to verify the consistency of the evaluation process.

Each university's concerned authorities examine the outcome and redefine the course if they believe it is worth achieving continuing excellence, i.e., Continuous Quality Improvement (CQI).

The curricula of Textile education have competitive depth with the curriculum of every other country, but breadth is not as competitive as depth. Therefore, all institutes now emphasize the curriculum to solve the dynamic problems of engineering (Fig. 5).

4.3 *Washington Accord*

The Washington Accord, founded in 1989, is an accredited training pact for undergraduate engineering programs. It acknowledges that the engineering programs approved by those signatories are substantially equivalent and that the other member countries accept graduates of accredited programs in each of the member countries as having fulfilled the academic needs [7]. Under this agreement, the university's mission statement must be very explicit and trickle down to the Educational Objective Program (PEO). Most stakeholders are involved. Industry, instructors, students, and graduates are considered stakeholders to formulate the educational goal of the initiative. The PEO is evaluated using the key output measure, and CQI processes (Continuous Quality Improvement) are ensured in each education domain.

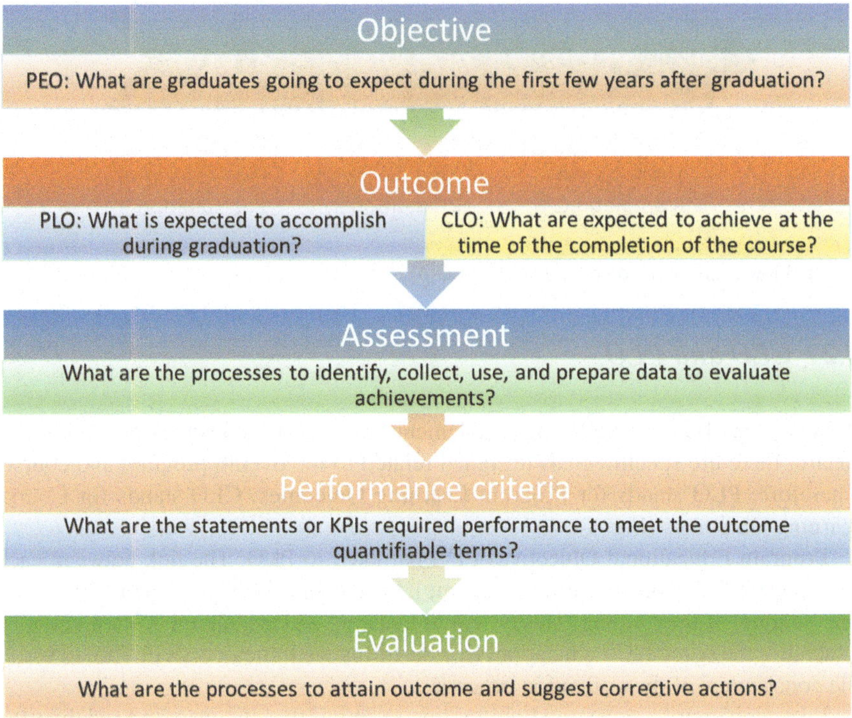

Fig. 5 PEO to PLO mapping [6]

4.4 Domains of Learning

We are lifelong learners as humans. We start learning when we are born and continue to learn throughout our lives. We continue to incorporate new information into what we already know as we gain new experiences (Fig. 6).

Learning, on the other hand, is more than just a mental (thinking) process. Attitudes, habits, and physical abilities can all be learned. These various categories create three learning domains.

The learning has been classified into three domains as per Bloom's Taxonomy, i.e., cognitive, (affective), emotional, and psychomotor, and assigns a hierarchy to each of these areas, corresponding to different levels of learning. All Pakistani Universities work their best by adopting modern techniques and strategies to prepare their students as excellent life-learners.

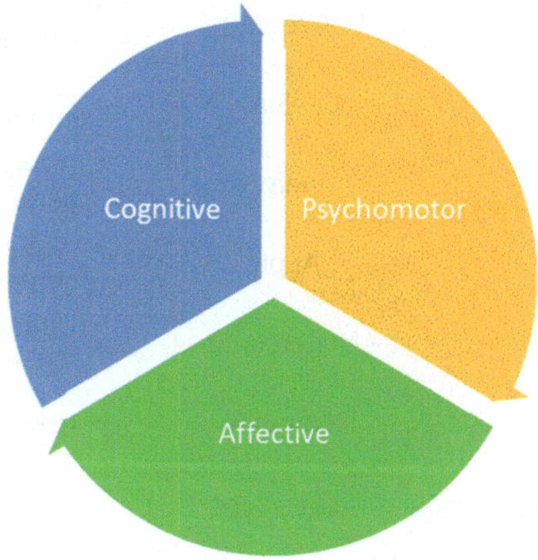

Fig. 6 Learning domain of OBE [8]

4.5 Knowledge Domain (Cognitive, Thinking)

Bloom's Taxonomy is a hierarchical structure that characterizes students' cognitive skills, ranging from retrieving knowledge that would be essential to assessing, including judging, and expressing an opinion on information. The knowledge domain idea suggests how much a student has perceived, understood, and discovered from the acquired education. The Bloom Taxonomy idea has been followed by Textile Universities of Pakistan, just as other world-class universities do worldwide, to evaluate the level of knowledge present among students. Bloom's taxonomy is important method teachers and educators use to build lesson plans and assessments to assist critical thinking [9]. The idea of Bloom Taxonomy is explained in Table 3 (Fig. 7).

Table 3 The idea of bloom taxonomy [10]

Level	Attribute	Explanation
Level 1	Remember	All about remembering and memorization
Level 2	Understand	Able to explain concepts
Level 3	Apply	Able to make the application of concept and knowledge
Level 4	Analyze	Able to differentiate, organize, relate, compare, contrast the application results
Level 5	Evaluate	Able to make the decision or defend what was analyzed
Level 6	Create	Able to produce new and original work

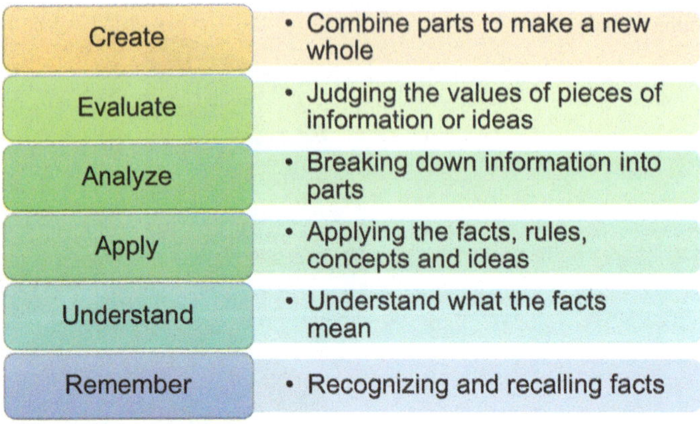

Fig. 7 Cognitive, thinking [11]

4.6 Skill Domain (Psychomotor, Doing)

One can never deny the importance of the presence of skills among students. Knowledge is useless if students do not have the skills to practice that knowledge in their respective fields, or it can be said that along with knowledge, the presence of skills is a complementary factor. Keeping the importance of skills in mind, Pakistani Textile Universities do their best to grow some essential skill set among their students. The areas of Psychomotor include resistance training, balance, and usage. Design of these abilities spectrum of disciplines and is calculated in tempo, accuracy, scope, processes, or execution techniques. Psychomotor skills vary from manual activities, such as digging a ditch or washing a vehicle, to more complicated projects like running a complicated piece of technology or performing. The table for Psychomotor is shown in Table 4.

4.7 Effective Attitude Domain (Feeling)

Apart from having excellent knowledge and unique skill sets, the other important thing is what attitude has developed among students. How they feel, what sort of mental outlook they possess, and how effectively they manage things professionally and personally. Educational institutes are not about where students come, learn books, go out and earn money. The core purpose of education is to build excellent creative and analytical human beings who make their society a better place to live in for themselves and other fellow beings. Pakistani Universities aim to build qualities in the practical attitude domain (mentioned in the table below) by believing that students will become skilled in any attributes of a practical domain if they want or need to do it. The attitude of the affective domain is obtained or created through various tasks

Table 4 Psychomotor level and its attributes [10]

Level	Attribute
Simple	*Perception*
	Distinguish, *Hear*, *See*, *Smell*, *Taste*, *Touch*
Simple	*Set*
	Adjust, *Approach*, *Locate*, *Place*, *Position*, *Prepare*
Moderate	*Guided Response*
	Copy, *Determine*, *Discover*, *Duplicate*, *Imitate*, *Inject*, *Repeat*
Moderate	*Mechanism*
	Adjust, *Build*, *Illustrate*, *indicate*, *Manipulate*, *Mix*, *setup*
Complex	*Adaptation*
	Adapt, *Build*, *Change*, *Develop*, *Supply*
Complex	*Organization*
	Construct, *Create*, *Design*, *Produce*

and methods inside and outside the university. The affective domain of learning can also be understood by the below-given Fig. 8.

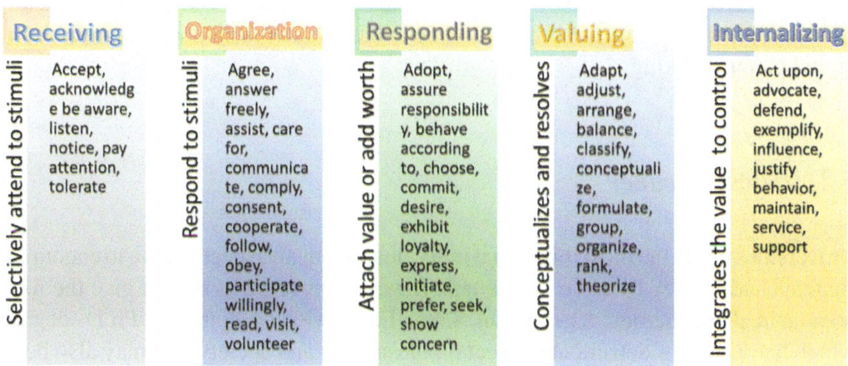

Fig. 8 Attributes and their definition with examples for effective attitude domain [10]

5 Learning Management System

A learning management system is a software application that manages educational courses, training programs, and learning and development programs through administration, documentation, tracking, reporting, automation, and delivery. The concept of a learning management system arose straight from e-Learning. All Textile Universities of Pakistan have an online portal over which they manage all educational activities. They are supposed to enter all the necessary details of each student such as marks, attendance, essential online material over their web portal. Students are assigned their accounts with complete details of their registered courses, mark sheets, and attendance details. Whereas teachers also have their online portals displaying the details regarding the courses which they are teaching. HOD (head of the department) can view each teacher's portal to track their performances.

6 Life-Long Learning

Universities are also attempting to cultivate the attribute of lifelong learning by using the following strategies.

6.1 Conferences and Training

Educational Institutes organize different conferences from time to time, such as Dyce, where people from across the area gather and share their latest textile and fashion market experiences and expertise. Training sessions are also conducted for faculty and students to improve the practical knowledge of such skills, such as training on digital pattern making.

6.2 Guest Speaker

Universities invite the most influential individuals from all aspects of life to encourage students and faculty to make life better. However, universities do not give the most popular real-life heroes, e.g., Abdul Sattar Edhi, with an honorary Ph.D. degree, which has a poor or detrimental effect on the system and society and may also be an obstacle to public acceptance of lifelong learning attitudes.

6.3 *Opportunities for Post Doctorate*

There are a very limited number of individuals who go to do the post-doctoral degree on their own. There are no opportunities to do a post-doctoral degree in Textiles or fashion at any university in Pakistan. There is no system for helping people to do post-doctoral studies at other universities. It may also be a factor for people not to get a professional development mindset.

7 Evaluation Methodologies

Students are assessed using both direct and indirect methods. Rubrics are used as a tool for assessing students.

7.1 *Rubrics*

Rubrics are a collection of rules used to assess students' activities (presentation and assignments). All the attributes that must be measured and their weighted %ages are listed in the rubrics.

7.2 *Direct Method of Assessment*

Methods of direct evaluation enable students to demonstrate expertise. That is, students must do something measurable or observable actively. Meaningful using the information and skills described in the SLO and outcomes Scoring processes are quantified, such as the correct %age, Total rubrics, or marks. There are examples listed below and skills and knowledge that precisely assess the anticipated achievement of the performance.

Attendance

To establish discipline, students must attend 75% of the lectures given to qualify for the exam to sit in the classes.

Seasonal Marks

Teachers evaluate students based on their interest in learning about the knowledge, developing skills, and the student attitude throughout the course. Such marks are added to the final examination.

Examination

Educational Institutes are adopting the semester structure in which students are met at the end of each semester by taking the mid-term and final term exams. Thus, the students study two semesters a year. To pass the course in the exam, students need to get 50% marks in the exam.

Open-Ended Lab

Except for a practical title, students in an open-ended lab works independently to complete the practical. Bloom's Taxonomy is used to identify learning levels for all courses. People that work in the industry are also involved in the course design.

Internships

In the sixth semester, students must complete a one-credit-hour internship. Some students are offered paid internships, although not all students are. Furthermore, not every university provides each student with an internship placement. The internship report is a direct way of assessment that students provide at the end of the internship. Feedback from the industry where students are interning is also considered, known as the indirect assessment method.

Research Project

To obtain an engineering degree, students must do a research project to demonstrate some creativity or analytical capabilities earned through the four-year curriculum. After completing research, they have to defend and present their work in front of HOD; instructors and sometimes external experts from the industry are also called to assess the research work of students. This research is considered an essential tool for students; it evaluates their performance and provides them an amazing opportunity to apply their knowledge practice and give them insights into what is going on globally.

7.3 Indirect Method of Assessment, i.e., Mentorship Program

Each student's knowledge, competence, and behavior are assessed using an indirect method of assessment that includes feedback from the industry, seniors, and the community. Mentorship is an indirect evaluation technique, but its primary goal is to instill confidence in students to engage with people currently working in the field and learn about its difficulties. Students at some universities, such as NTU, can be mentored by seniors. Although this program is excellent in student development, NTU may need to rethink how it is managed. Unfortunately, every university does not offer a mentorship program to its students.

8 Admission of Students

Students must also have a fundamental awareness of textile and arts to get a place to study textiles in Pakistan. Students must pass the basic textile test with at least 60% to do an engineering program. Universities also administer an entrance examination system. Students with more marks get a chance to do engineering in their chosen discipline by adding F.Sc and entry exams. Students who like to study Textile Marketing and Management generally have a background in the industry, but there are no university limitations that students need to know those subjects. Admission is awarded based on academic qualifications (which is equivalent to F.Sc or A level). Students interested in fashion design or textile design are admitted to the F.Sc or A level of schooling, respectively.

9 Resources for Students

9.1 Computer Labs and Library

All Universities have their libraries. Students and professors at other universities can use the Virtual HEC library to get current books and articles. Each university has a 1 1 computer lab student-to-computer ratio for each class.

9.2 Career Counselor

Pakistani Universities have very restricted facilities for advising students on various career opportunities. The class counselor may take the initiative to support the students. However, at the time of writing, there was no career counselor office at universities.

9.3 Class Advisor

Each class has an advisor responsible for assisting the student with the problem of life each day. On the office door, the class advisor availability period is shown.

9.4 Accommodation

In addition to the mess and canteen, most universities have accommodation facilities, but not every university provides students with hostels. Therefore, students ought to pay for the accommodation and the mess.

9.5 Transportation

Universities typically provide transportation for students and staff to get around the city. Students who use public transportation are required to pay for it.

9.6 Health and Safety Arrangements

Universities have put the evacuation plan on walls, and in case of disaster, more students and staff are prepared for evacuation. Universities also uphold university hygiene standards for the health and safety of all. Finally, universities seek to act against offenders by enforcing the university's disciplinary law if someone uses illegal drugs or is involved in other illegal activities.

9.7 Scholarship Program

Students can get a need base and merit base scholarship required, but these scholarships are very limited. Not every student gets it, and not every student deserves it. Students are struggling a lot to handle the expenses. HEC has also intervened in the situation and is seeking to support students who face financial challenges.

9.8 HEC Scholarships

HEC has two types of scholarships for students and staff, i.e., National Scholarship (refer, Table 5) and Foreign Scholarship and Table 6.

Most international scholarships are now closed for doing a master's and Ph.D. when writing the article. However, under the Faculty Development Initiative, NTU will send some of its faculty members abroad in 2020.

Table 5 Detail of national scholarships

Name of scholarship	Eligibility
Scholarship for FATA and Baluchistan student Phase 2	Students must belong to FATA or Baluchistan Students have 16 years of education with at least 60% marks
Ehsaas Program	Newly admitted students The student has admission to the institute Degree Program allowed by HEC
Gwadar China Scholarship	Applicant must belong to district Gwadar,12 years of formal education with 50% marks, maximum age is 22 years
Prime Minister's fee reimbursement scheme for less developed area	Students belong to the specific area under the scheme Students who enrolled on merit on Masters/MPhil on the geographical territory of their province/Region/area Students enrolled in Ph.D. anywhere in Pakistan
HEC need-based scholarship	Any student whose family is financially struggling can apply
OGDCL need-based scholarship	The student must be enrolled at Mehran University. Students belong to a remote and vulnerable area of OGDCL operations/concessions of Pakistan
USAID funded merit and need-based scholarship	The student must be newly enrolled at Mehran University of Engineering and Technology or the University of Agriculture; also, the student must be struggling financially
Indigenous Ph.D. Fellowship Program	Pakistani/AJK nationals, Candidate must have 3 CGPA, OR no 3rd division in the annual system, maximum age is 40 years for faculty of public sector university and 35 for all others. The last date for submission of the application is April 18
Indigenous Scholarship Aghaz-E-Haqooq–E-Balochistan	Students belong to Baluchistan; Applicant must need to secure 50% marks in the HEC test. Along with no 3rd division is acceptable or have 3 CGP out of 4. The maximum age is 40 years for full-time regular faculty members and 35 years for all others

Table 6 Detail of foreign scholarship [12]

Scholarship name	Eligibility criteria
Hungary scholarship for Pakistani student	Pakistani and AJK nationals, Maximum age 35 years for master program, 40 years for Ph.D. who are serving public sector universities

9.9 Universities and HEC Collaboration for Doing Ph.D.

Students will go abroad for six months as part of the Ph.D. program under the international research support initiative program.

9.10 Research Grants

Research funding is also accessible on the HEC website under the R&D category. The most common research grants are competitive research grants, research support grants, mobility awards, and research for creativity grants [10].

9.11 Feedback of Ph.D. Students and Researchers

Students and researchers who have earned Ph.D. scholarships from HEC claim that the process is a little long, complex, and complicated.

10 Faculties for Textile Education and Faculty Development Program

The student to faculty ratio is 25 1, but preferably, universities are aiming to have a 20 1 ratio. The minimum faculty qualification is postgraduate, but preferences are given to individuals with a Ph.D. degree [10]. Universities such as NTU give further education to the faculty in various countries. There will be 10 NTU lecturers doing PH.D. In various countries in 2020, the NTU must be served for at least five years under the terms. Not every university provides workers this form of opportunity [13].

11 Industrial and Educational Liaisons

The textile industry has a strong relationship with NTU, and as a result, many textile graduates receive research funding from the industry, and students receive paid internships and even jobs before finishing their education. NTU also has influential industrialists on its governing board or executive committee. Furthermore, the All Pakistan Textile Mills Association (APTMA) has supported NTU in giving funding. However, few universities' faculties actively give industry consulting; institutions also collaborate before establishing courses. Therefore, all universities should have the same design and amount of industry liaison.

12 Extra-Curricular Activities

Extra-Curriculum practices can develop or enhance physical and mental health, develop time management skills, improve a wider variety of studies and social skills. For example, to strengthen their deficiency in some skills, students can choose sports. Sports at universities are available, but not every university has the same facilities. The linkage of skills and extra-curricular activities is given in Table 7.

Students can engage in various events, such as indoor sports, outdoor sports, literary activities. There are no facilities for playing or practicing in water sports such as swimming, surfing, skiing, boating, fishing, and rowing for outdoor sports students to play cricket, football, hockey, basketball, volleyball. The universities do not have a long tennis court. The University Sports Committees organize Intra University and Inter-University Sports Competitions. Not only at universities but even on the streets of Pakistan, cricket is the most common sport. There are only table tennis and men's gym in indoor sports colleges, and there are no other indoor facilities in universities. Pakistan's traditional sports such as Lattu, Punjabi Khabbdi, Pehlwani, Bull Karah are not popular university sports, but individuals love to play in different country regions. Fighter kites are a common sport at universities and in the region, but it is a dangerous sport; the government now bans it.

University students organize literary societies, publish news articles (covering university news) into Urdu and English literary magazines, and organize debates and poetry competitions between universities. The industrial magazine (TEX TECH magazine name) is published by some universities such as NTU, in which industry professionals, academic professionals, and activities write articles on textile industry issues in every aspect. Tax Tech does not have any influencing factor at all.

Once a year, the Rotary Club managed to get blood donations from university students (it is a student preference if they choose to donate) to serve patients who desperately need blood to save lives. Each department's societies also allow students to learn how to manage various events and be organized in real life.

Table 7 Table for linkage of skills and extra-curricular activities

Attributes	Extra-curricular activities
Improving general knowledge, and memorization skills	Puzzles, fun quizzes, scribes, and cards
Time management skill	Work for university magazine and newsletter, Work at different societies
Social Skills	Work at Rota tact Club, Get together, concerts
Improve physical and mental health	Indoor, outdoor sports
Improve literature skills	Poetry and debate Competition

13 Conclusions and Future Recommendations

The textile industry regulates Pakistan's economy, so education must endure a competitive edge in the global market. Pakistan's textile industry of product mix is a solid contribution for the international market, but textile universities need to develop education in advance to gain market share. In Pakistan, numerous textile universities follow the result-based education system by taking all aspects of the industry and student needs into account.

Universities must try to help the textile industry to get interested in adding value. Within and outside the country, students at universities should have the opportunity to share their work and thoughts. Both within the region and outside the region, universities will need to focus on student exchange programs. Universities should carry out collaborations with foreign brands. Universities should enable students to do business on their own. To achieve sustainable fashion and sustainable people, fashion companies, suppliers, and universities must work together. Finance firms must also assist and inspire students to grow their fashion industry.

References

1. Pakistan Commissioner's Organization Government of Pakistan [Online]. Available https://www.tco.com.pk/publications.aspx
2. https://www.psx.com.pk/psx/resources-and-tools/listings/listed-companies [Online]. Available https://www.psx.com.pk/psx/resources-and-tools/listings/listed-companies
3. Trade Development Authority of Pakistan [Online]. Available https://tdap.gov.pk/trade-statistics/#
4. Textile Industry in Pakistan, 18 02 2021 [Online]. Available https://en.wikipedia.org/wiki/Textile_industry_in_Pakistan#Significance
5. Davis MH (2015) Educational strategies outcome-based education. J Vet Med Educ 30:258–263
6. Department of Automotive and Marine Engineering (2018) [Online]. Available https://atd.neduet.edu.pk/OBE
7. Khalid Mahmood KSK, Kiani S (2015) Implementation of outcome based education in Pakistan A step towards Washington Accord. In IEEE 7th international conference on engineering education (ICEED), Kanazawa, Japan
8. Domains of Learning (2021) [Online]. Available https://thepeakperformancecenter.com/educational-learning/learning/process/domains-of-learning/
9. Revised Bloom's Taxonomy Action Verbs [Online]. Available https://www.apu.edu/live_data/files/333/blooms_taxonomy_action_verbs.pdf
10. Pakistan Engineering Council [Online]. Available https://pec.org.pk/downloads/Accreditation/2/a/5.%20Program%20Evaluation%20Guidelines%20for%20PEVs.pdf
11. University of ARKANASAS [Online]. Available https://tips.uark.edu/
12. Higher Education Commission, Pakistan [Online]. Available https://www.hec.gov.pk/Pages/PageNotFoundError.aspx?requestUrl=https//www.hec.gov.pk/english/scholarshipsgrants/Pages/NationalScholarships.aspxac
13. National Textile University [Online]. Available https://ntu.edu.pk/notice-details.php?id=853

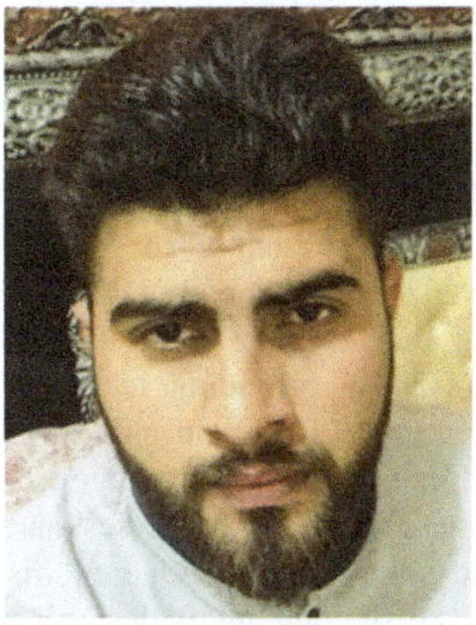

Gohar Ali Hayat Gohar Ali Hayat is currently working as a lecturer in Fashion Design at the National Textile University Karachi campus. He had done his graduation in Fashion Design and Technology from National Textile University Faisalabad in 2017. Due to his remarkable achievements, he is considered as Gold Medalist in graduation. He completed a post-graduation in Art and Design from Wuhan Textile University China in 2019. He is an enrolled Ph.D. Scholar at Soochow University China.

Muzammal Hussain Muzammal Hussain did graduation in Textile Engineering from National Textile University Pakistan. Later, he did his post-graduation in Clothing Management from Heriot-Watt University UK. Engr. Muzammal worked at TESCO's clothing brand F & F in the UK, Arcadia's brand Burton in the UK, and Gulistan group of companies Pakistan. Currently, he is a lecturer at the National Textile University Karachi campus.

Muhammad Qamar Khan Muhammad Qamar Khan is currently serving as Chairman of the Department 0f Textile and Clothing. He has done graduation in Textile Engineering with a specialization in garments manufacturing in 2013 from National Textile University Faisalabad. Dr. Qamar pursued his post-graduation and Ph.D. from Shinshu University Nagano, Japan. In 2019 he rejoined National Textile University Karachi Campus as Chairman of Textile and Clothing of the department. He has more than 40 impact factor research publications and six book chapters publications.

Zafar Javed Zafar Javed pursued his doctorate. from the Tampere University of Technology, Finland, and post-graduation in Advanced Textiles. from Ghent University, Belgium, and graduation in Textile Engineering from National Textile University Faisalabad. He is a Textile Engineer and holds the position of Faculty Director of School of Arts & Design at National Textile University, Faisalabad, Pakistan. His prime expertise is in teaching and research at the university level. Dr. Zafar is also involved in providing technical assistance to textile ventures in Pakistan. The main area is to combine Textile Design, Fashion Design, artificial intelligence, and Apparel Manufacturing to create efficient processes and the new breed of sustainable and smart textiles. One of his latest consulting projects with a Swedish company is designing and developing a genuinely sustainable and honest apparel production unit. In this venture, a production unit has been set up from scratch, and a blockchain-based, transparency-inducted sustainable clothing brand has been launched and is now a successful enlisted company in Sweden with its production base in Pakistan. He has also authored several research papers and international conference presentations from technical and processes management aspects of the Apparel and Fashion areas.

Prospects of Textile Education in Afghanistan

Masaood Moahid and Ghulam Dastgir Khan

Abstract The textile industry has long been regarded as a critical enabler of industrialization in developing countries. This chapter investigates Afghanistan's textile industry and discusses the prospects and significance of textile education as a crucial success factor in promoting the industry. Afghanistan's textile was one of the most well-known manufacturing industries, with a long history that blossomed in the 1970s. Before the wars and conflicts in the country, textile-related operations such as cotton and wool production, manufacturing, sales, and marketing were critical economic activities in Afghanistan. However, like other industries, the textile industry was seriously affected by the war and conflict. The restoration and development of Afghanistan's textile industry can generate substantial revenue for the government and numerous job opportunities for Afghans. Foreign competitors, primarily from Asian countries like China, are currently putting pressure on the fledgling textile industry. The Afghan textile industry must carefully examine its business models and take the necessary steps. Another significant impediment is a lack of technological expertise and skills. On the other hand, Afghanistan provides favorable conditions for the textile industry due to comparatively low labor costs, low-cost factory space, and potentially significant agricultural raw materials and resources. Cotton, cashmere, silk, fur, skin, and hides are all available as home raw materials. Policies promoting the production of these products can have a significant impact on people's livelihoods. Most importantly, textile education can have a substantial effect on the sector's growth. As a result, careful consideration of these critical aspects by the Afghan government and other stakeholders would significantly impact reviving the country's crippled textile industry.

Keywords Textile · Industry · Education · Afghanistan

M. Moahid (✉)
Faculty of Agriculture, Nangarhar University, Jalalabad City, Nangarhar 2601, Afghanistan

G. Dastgir Khan
Graduate School for International Development and Cooperation, Hiroshima University, 1-5-1 Kagamiyama, Higashi-Hiroshima, Hiroshima-ken 739-8529, Japan

X. Yan et al. (eds.), *Textile and Fashion Education Internationalization*, Textile Science and Clothing Technology, https://doi.org/10.1007/978-981-16-8854-6_5

1 Historical Perspective and Background of the Textile Industry in Afghanistan

The textile sector has been regarded as an essential facilitator for developing countries' industrialization. Because of its lower fixed costs and labor-intensive nature have played a critical role in many countries' growth and industrial advancement. As a result, it is often a leader field for developing nations seeking to industrialize [10].

The textile industry in Afghanistan was one of the most well-known manufacturing industries, with a long history of several decades. The industry flowered in the country in the 1970s when the Afghan government built mega-textile production plants in the western, northern, and southern parts of Afghanistan. These major plants were then built for domestic and foreign markets, whereby Afghanistan would generate huge amounts of foreign currency. The country would export its textiles to and from the region centuries ago [2].

Textile-related operations, such as cotton and wool production, manufacturing, sales, and marketing, were important economic activities in Afghanistan prior to the wars and conflicts. They provided livelihood opportunities for many rural and urban Afghans. For example, Lin cotton was manufactured in the country during those years [6]. However, owing to war and turmoil, the country's textile industry, like other industries, saw a vertical and horizontal downturn. Cotton cultivation peaked in the early 1980s at about 350,000 tons per year on 188,700 ha of soil. In 2003, the estimated yield was only 30,000 t from 46,960 ha [2].

The textile industry is split into two subsectors in Afghanistan: organized and unorganized. Spinning, clothing, and apparels are part of the integrated industry, which involves modern machinery and techniques. In the development of fabric and clothing, the unorganized textile subsector mainly uses conventional woven or spun processes. This subsector is an important part of the textile industry, and it employs a significant portion of the country's labor force. Even today, it makes a significant contribution to jobs and the economy of the country.

A tradition of spinning and weaving has been passed down over the centuries [10]. Many Afghans have followed this centuries-old practice, which has become a source of income for many households. Many parts of the country have artistic instincts when making textiles such as clothes, hats, scarves, rugs, waistcoats, and other products. Many small and medium enterprises operate throughout the country, producing various textile products for domestic and international markets.

Given its labor-intensive nature, the textile industry has focused on poverty reduction and economic growth since establishing a new government and transitioning to a market economy in 2002. Accordingly, the government and its donor partners have begun programs supporting private cotton farmers, ginneries, and textile mills since 2002. Nonetheless, the programs and projects were on a small scale and fragmented manner. As a result, the sector has yet to grow to its full potential in the country. Moreover, among several challenges to the sector, lack of skills and educational programs dramatically slows the textile sector's development.

Textile education, which includes all fiber-based artistic practices that include textile materials, such as apparel and fashion design, costume, home furnishings, and textile art, has the greatest effect on the sector's growth, as per the experiences of other countries. The Afghan government and policymakers need to be aware of the textile industry's employment potentials and options available. Professional education and skills in textile design, fashion, and art, on the other hand, must be promoted. By receiving an adequate and appropriate education, farmers, textile workers, and designers learn textile-related materials processing and the whole textile design process.

Despite this, the Afghan textile industry seems to have a bright future buoyed by high domestic and export demand and its raw material production potential. Furthermore, in Afghanistan, the retail sector has grown rapidly in the last two decades due to rising consumerism and disposable income. Through their entrance into the Afghan market, some foreign players have also lifted the bar. Thus, the textile sector is expected to have positive growth prospects with stable input supplies, textile education, adequate capacity utilization, and steady domestic demand.

No formal public and private institutions are providing technical and professional education and skills in the textile industry. According to the industry stakeholders, a textile diploma can teach how to utilize textures, patterns, and color through experimentation with printing, dying, manipulation, and embellishment techniques. Students looking to foray into this field will understand Afghan fabrics, their surface ornamentations, and their role in the global industry. Furthermore, active cooperation between indigenous designers and partners in this segment will ensure that students will have credible, technical experience to address both local and foreign markets by the end of the course.

Cotton, cashmere, linen, and silk processing are all crucial textile-related commodities. Farmers in the country are eager to adopt and develop these products, critical in the textile industry. Therefore, the promotion of these commodities' output could significantly impact the country's textile industry growth. In this regard, research and development activities are needed.

This chapter mainly reflects on the prospects of textile education and its importance in developing Afghanistan's textile industry. The following part of this chapter focuses on the country's vital textile-related commodity production. Following that, the prospects of the Afghan textile industry are discussed. The prospects for textile education in Afghanistan are included in the subsequent parts. Finally, the chapter concludes with a summary and recommendations.

2 Textile Related Commodities' Production and Processing in Afghanistan

2.1 Cotton Production and Processing

Afghanistan is a rugged and mountainous country with only 12% arable land. Despite this, agriculture is critical to the survival of the people. Agriculture provides a livelihood for about 80% of Afghans [5]. It has traditionally been a major driver of the Afghan economy. Afghanistan used to provide enough food and non-food agricultural products for domestic use before the decades of war. Besides, the country exported various agricultural products, including almonds, pomegranates, pistachios, raisins, apricots, poultry, and cotton. However, like other sectors, the agriculture sector was seriously damaged by the war and conflicts between 1980 and 2002. Furthermore, only six percent of 12% of cultivable Afghan land is currently under cultivation [13].

Nonetheless, the country's agriculture is now on the mend, thanks largely to international aid. Nevertheless, the sector continues to be the primary source of income for most households. Agriculture employs 38.3% of the workforce and accounts for about 24% of the country's gross domestic product (GDP) [19].

Cotton is one of Afghanistan's most important crops for fiber processing, grown on approximately 33,000 ha of land with a total yield of 42,872 tons and productivity of 1299 kg per ha in 2016 [15]. Furthermore, as a byproduct of cotton cultivation, the cotton cake is used as animal feed, while its seeds make edible oil.

The establishment of the Northern Province joint-stock company in 1935 was the first step toward improving cotton cultivation in Afghanistan. The corporation put in a five-million-dollar investment at first. In Afghanistan's northern area, the company began a two-pronged campaign to advance cotton cultivation and textile production [15]. First, via the establishment of textile factories, the company's primary goal was to encourage cotton production, increase exports, and reduce import dependency.

The Northern Provinces were chosen because the Afghan royal government was in the process of reclaiming large swaths of uncultivated land by unpaid labor at the time. After cultivating the land, it was sold to the general public for a low price. Following that, the company acquired a large amount of land and took in many farmers from all over the country to begin land cultivation. Farmers were expected to plant cotton on at least 6% of their property. In addition, the firm began promoting American long-fiber in the area. Short-fiber cotton was then used instead of the more suitable long-fiber cotton.

After completing two cotton processing factories in 1937 and 1942 at Jabal Seraj and Pole-Khomri, respectively. As a result, cotton became known as Afghanistan's "white gold", and it became a vital export for the country, accounting for 14.3% of overall export earnings in 1940, compared to 0.9% in 1937 [2].

The ginning and seed processing plants were located near cotton cultivation centers in rural areas. Activities such as spinning and weaving were carried out in urban areas. Five government-owned firms dominated the textile industry, which was seen as the most important market. Between 1960 and 1980, Afghanistan used to own many larger textile firms in Balkh, Kabul, Baghlan, Kandahar, and Parwan, employing about 30,000 people [11].

However, over the past few decades, the market has declined by manifold. The shortage of manufacturing plants is the greatest obstacle to the country's cotton crop. In addition, several production constraints in the country contribute to low cotton yields, including the use of sub-optimal inputs, especially fertilizer [11, 16].

The Afghan manufacturing sector is also in an early growth period. Compared to the country's scale and capital endowment, its contribution to domestic jobs and production is minor. The industry sector is relying on small-scale manufacturing, primarily textiles and mines. The sector in the country accounted for 10% of the labor force, and its value-added accounts for 22.7% of GDP. Figures 1 and 2 reveal the proportions of employment and percentages of value-added in various sectors for the years 2008–2016. Furthermore, the sector is the only one that employs most women (65% of all manufacturing workers are female) [18].

Manufacturers recently lobbied for guidance on restoring Afghan textiles. Revitalizing this valuable sector of the industry could generate country-wide economic growth. The country is endowed with high-quality cotton, silk, and cashmere. As a result, it has many opportunities for the textile industry to develop, a strong source of jobs and livelihood. Afghan President Ashraf Ghani recently opened a seven million USD cotton processing plant in Herat province's industrial park. The Afghan White Gold Cotton Industry factory is one of 30 new factories recently opened in the park. According to the interviewed textile industry stakeholders, it buys 20 tons of cotton a day from local farmers.

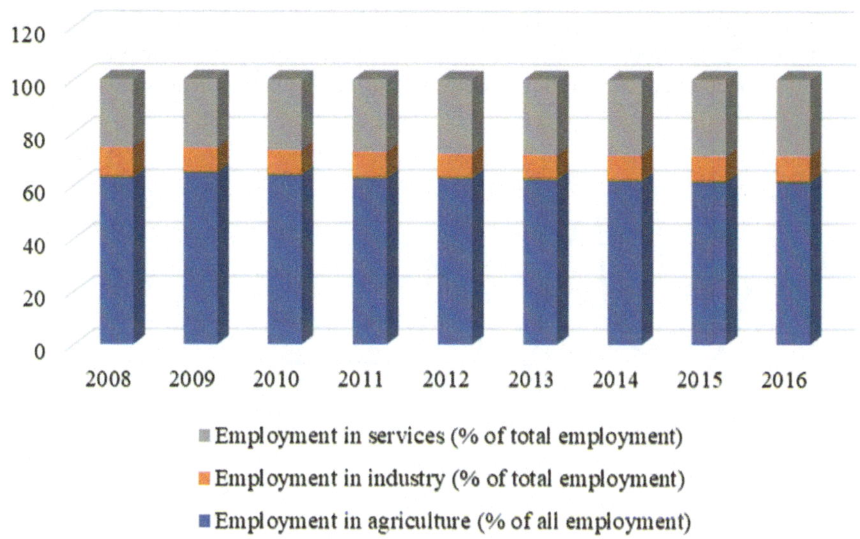

Fig. 1 Sectoral employment in Afghanistan 2008–2016. *Source* World Bank. Online at: http://databank.worldbank.org/data/source/world-development-indicators

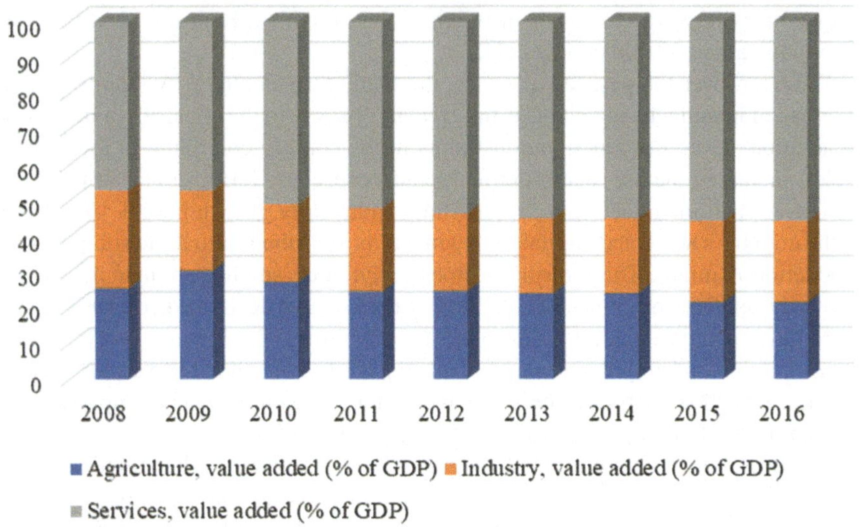

Fig. 2 Sectoral value-added in Afghanistan, 2008–2016. *Source* World Bank. Online at: http://databank.worldbank.org/data/source/world-development-indicators

Cotton production, harvested area, and exports are depicted in Fig. 3. Cotton output in the country fluctuated a lot between 1960 and 1988. It was raised between 1960 and 1964, owing primarily to the irrigation of new fields. In 1960, gross demand was about 80,000 bales, but by 1963, it had risen to 175,000 bales.

The production was extremely unpredictable between 1963 and 1973. However, it increased by 253,000 bales in 1983. Thanks to large extension services and textile sector enhancement policy of the government during that period. The yield improved

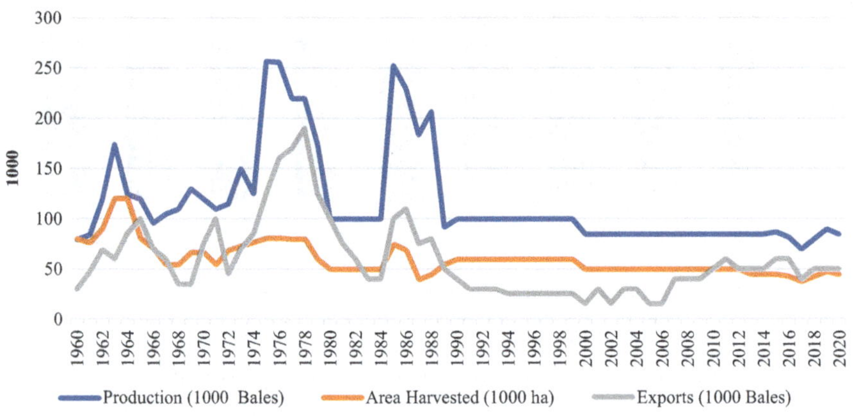

Fig. 3 Afghanistan's cotton production, area harvested, and cotton exports between 1960 and 2020 (*Source* NSIA [14])

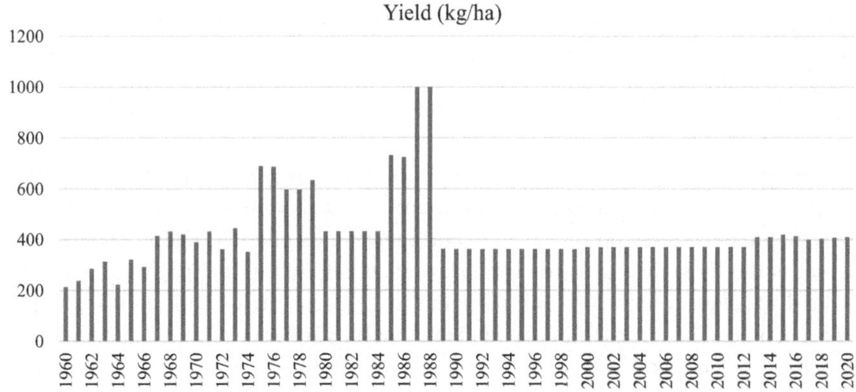

Fig. 4 Per hectare yield of cotton in Afghanistan between 1960 and 2020 (*Source* NSIA [14])

dramatically in 1975 (Fig. 4). Between 1982 and 1989, the Soviet-backed socialist government's major sources of income were cotton cultivation, manufacturing, and exports.

The average yield increased to one metric ton per hectare during this period (Fig. 4). The per hectare yield was highly volatile between 1975 and 1988. It significantly rose between 1982 and 1988 mainly due to supportive government policy. There has not been any significant change in the cotton yield from 2002 onwards.

While other commodities such as wheat production rose after the country's new government took office in 2001, cotton production remained low. On the one hand, this shows the failure and devastation of government-owned textile manufacturing factories between 1992 and 2001, when the country was at war. However, on the other hand, it reveals that the current efforts to promote cotton production have been unsuccessful.

2.2 Afghan Silk

The northwestern Afghan city, Herat's long history of silk production lies in Silk Road. The country presently undergoes a recovery following years of decline in silk production. Silkworms make fiber of silk. Mulberry plants, the food of the silkworms, is familiar to the producers in Herat province [5].

Sericulture in the Western Herat province of Afghanistan is part of agriculture that has expanded, but similar to cotton cultivation, persistent conflicts and wars have drastically stagnated. However, Afghanistan is now trying to restore sericulture and produce silk in the province to revitalize the silk trade between its western region and the countries in the region, particularly neighboring China.

Sericulture was part of local people since the industry had become popular in Zindajan, a district in Herat, predominantly an agricultural area.

Through its projects, the Afghan Ministry of Agriculture imports silkworms from China kept in separate, secure boxes and distributed them to farmers and cocoon producers in the district annually. As a result, hundreds of families in the district are now engaged in silkworm breeding, which has been booming recently, more so than in the previous decade.

Women are primarily involved in keeping silkworms at home in Herat province. They are fed mulberry leaves. As the amount of silk production grows, the government should encourage producers to export their goods internationally. The Chinese market is the most desirable for Herat province silk due to its proximity and high demand. It takes at least a month for them to get fresh, active silkworms ready to make cocoons. The thread is then wound into bobbins and transported to other parts of Afghanistan, including Mazar, Afghanistan's fourth-largest city and a vital commercial center. According to the agriculture department officials in Herat, farmers in the province have produced tens of tons of cocoons in 2016, which is a significant increase over the previous three years.

The government should help producers export their goods overseas as silk production increases. The Chinese market is very appropriate for exporting silk produced in Herat. Silk producers in the district say that delicate fabrics, beautiful clothes, and high-quality colorful headscarves would be manufactured by the Herat craftsmen from raw silk and exported to foreign countries, particularly China. The price of one kg of raw silk at a local market in Herat, according to locals, is 240 USD, and if exported to China, the price could be even higher.

Cocoon farmers can receive technical training and regular silkworm kits as part of the government's cocoon and sericulture initiative. The Afghan government is committed to helping the industry develop and expand into new markets, especially in China. As the country's silk production grows, the government should encourage the industry to expand and export its goods abroad. The Chinese market is the most desirable for Herat province silk due to its proximity and high demand.

2.3 Afghan Cashmere

Despite ongoing military conflict and extreme poverty, Afghanistan is the world's third-largest cashmere producer. Mongolia's second most vigorous producer generates 15% of world cashmere, well behind China, at 70%. The country exports about 1000 tons of cashmere per year, accounting for about 7% of global demand [9]. Thus, cashmere processing in Afghanistan has much promise. However, there is a need for services and schemes to provide money for rural citizens to run profitable and long-term ventures.

The cashmere goat is Afghanistan's, Turkey's, Iraq's, Iran's, Mongolia's, and China's domestic animals. It takes a goat around a year to grow enough wool to make only one scarf. Farmers in northwestern Afghanistan have used the dense undercoat shed from goats every year in the spring season as a source of fuel to cook food and heat their homes for centuries. However, the Afghans have only recently discovered that this commodity could be spun into more profitable goods. Buyers come from many countries to purchase cashmere, with each animal yielding up to 250 g a year.

According to [19] estimates, Afghanistan has seven million goats capable of producing cashmere. However, only 30% of them are trimmed for cashmere. Chinese merchants buy the majority of the cashmere and sell it to garment factories. Recently, the Afghan government developed a cashmere development framework to recognize this lucrative demand aimed at the high ends of the cashmere market, where only one sweater will cost up to USD 1000. This strategy forms part of the efforts by the Afghan government to revive the textile industry in the country.

3 Prospects of the Textile Sector of Afghanistan

3.1 Trade in Textile and Clothing Products in Afghanistan

The imported value-added goods win almost the entire market share inside the country, which is a great challenge to domestic companies. In addition, the local products are considered poor quality as they are produced using outmoded technologies and manufacturing methods. Afghanistan imports more than USD 500 million worth of textiles per year from China, Iran, Turkey, Pakistan, India, and South Korea. After wheat, gasoline, and cooking oil, it was the fourth-highest commodity imported to Afghanistan in 2017, 2018, and 2019 [8]. In 2016, Chinese textiles had a share of 39% compared to the overall import, and by AFN 4.1 billion, the textile imports rose compared to last year (18.02%). On the other hand, textile products constituted only 1.7% of Afghanistan's total exports. As a result, the total amount of textile export was only 11.9 million USD in 2019 [8].

Textile manufacturing firms can make a final product using less effort on cotton, silk, and wool. However, Afghan manufacturers export their products in unfinished form. They process cotton and then export it to Pakistan and Iran to make fiber out of it. The same cotton fiber is then imported back in finish form at a high price for textile manufacturing. Moreover, as imported value-added products capture the most market share, domestic firms are a major threat. A high portion of allocated savings for the textile and clothing of an individual is spent on these products. The general perception about the local product is of low quality due to old technology and production techniques. Figure 5 reveals the textile imports of Afghanistan from 2011 to 2019.

Fig. 5 Textile imports of Afghanistan from 2011 to 2019 (*Source* NSIA [14])

3.2 Growing Economy and Domestic Market

The textile and cloths expenditure makes a huge part of the people's income being a developing nation. Therefore, the country's textile industry has enormous promise. Afghanistan's GDP per capita growth has risen from USD 207 in 2002 to USD 647 in 2020 [8]. Even though a relatively stable price increase is seen from 2002 to 2020, this implies increasing purchasing power for citizens to buy high-quality textile products and help drive growth in the country's textile industry. In addition, faster economic growth beginning in the early 2000s has led to the emergence of an expanding middle class of millions of consumers, with the capacity and propensity to purchase higher-priced items.

Nonetheless, most Afghan consumers are highly price sensitive. The average Afghan household spends about 60% of its income on food and, as a result, pays discretionary income carefully. Price-sensitive, lower-income households account for the bulk of more than 30 million Afghan consumers [8]. Therefore, production efficiency and declining real prices for yarns and textiles can stimulate growth in demand for textile products since the early 2000s, particularly those made of artificial fibers.

3.3 Regional Cooperation

The Afghan government has to implement important collaborative initiatives with friendly countries to restore the textile industry. The collaboration should aim to enable the country to substitute imports and increase its textile exports. The collaborations can also help the textile sector of the country to employ advanced technology for textile manufacturing.

Uzbekistan and India have shown interest in Afghanistan's textile sector development among the countries in the region. Therefore, cooperation with Uzbekistan in the textile industry can undoubtedly benefit Afghanistan in various ways.

Uzbekistan is a rapidly growing vertically integrated cotton-producing country, in which cotton is processed from seed to finished product inside the country. This allows for cost savings in logistics, which impacts the product's price and global competitiveness. This is what, which is the most important experience that needs to be shared with Afghan textile industry stakeholders.

In 2016, experts from the Uzbekistan textile industry visited Afghanistan to research the country's textile industry and make recommendations. The Uzbek experts visited cotton processing plants and textile mills in Afghanistan. The Uzbek delegation also signed a memorandum of cooperation with Afghan officials in Kabul to develop the sector. The parties concluded that Uzbekistan would provide Afghanistan with consultations and technical assistance.

Through the agreement, Uzbekistan has also consented to assist Afghan businesses in developing business strategies and organizing educational courses for Afghan specialists at an Uzbek university, among other things.

India has also promised to help improve the Afghan textile industry, including capacity building, research, technological cooperation in product development and manufacturing, research and certification, and joint trade missions for mutual collaboration.

Joint ventures with foreign companies can also have crucial benefits for the sector's development. A synergy result raised from such joint ventures can be a good tool to achieve high profits from the Afghan market and exports. Local investors can avail of modern technology, production techniques, quality management, and cost reduction techniques from the countries with well-integrated advanced textile sectors. While at the same time, the local market and distribution channels' understanding of the Afghan market can produce a synergy effect.

3.4 The Need for Textile Education and Skills

One of the most important performance factors in the textile industry is having essential leadership skills and expertise. Productivity gains can be achieved by improving management skills and delivering adequate training to employees in larger companies. Mismanagement and a lack of skills are the most common causes of textile business failure. Therefore, education and skills development are relevant to the textile sector's growth [3]. Academic institutions cannot produce graduates who move to industry careers and contribute positively to the Afghan economy, which is a source of concern. Therefore, universities that prepare graduates for viable career prospects should be encouraged. As a result, it is time to consider how to fit educational opportunities into relevant and viable industries and revise educational programs to account for potential growth opportunities.

The current textile factories in the country rely on foreign employees. For example, the Aryana Balkh factory, located in the northern province of Balkh, relies on Pakistani skilled and semi-skilled labor. This raises costs and causes a halt in production.

Afghan employees were unable to operate the equipment and perform the necessary tasks efficiently without having foreign employees. Recruiting local labor lowers costs, but it requires an initial investment in educating them. Currently, some factories also hire Pakistani consultants to train local workers. Thus, education, expertise, and skills are essential for the sector's development.

Furthermore, financial expertise, company legal knowledge, and logistics are critical in managing Afghan textile businesses. In this industry, management skills are in high demand at all levels. Knowledge of life, history, and the local textile sector should be promoted to achieve higher managerial skills and more experience in the textile industry. The government and other stakeholders should provide education and training programs, seminars, scholarships, and study abroad opportunities, among other things, to teach and upgrade the managerial skills level of local employees.

Therefore, Afghan textile workers with technical skills such as information, experience, innovative thinking, and creativity will be prepared to handle textile firms independently, lowering the high labor costs linked with foreigners' recruitment. In addition, the textile industry can stand to benefit by introducing evolving and continuous advances, and managerial skills and expertise are at the heart of this process.

3.5 The Textile-Related Commodities Production Potential

The potential for cotton production in Afghanistan can be understood from the production figures reported in 1983, in which Afghanistan produced about 350,000 t of cotton. The quality and potential volume of cotton produced in Afghanistan are in demand in the international market at attractive prices to Afghan producers and ginners.

In particular, Kunduz and Helmand's vast land and farmers' acquaintance with cotton cultivation can provide valuable advantages to allow the cotton industry to grow. However, exporting cotton produced in Afghanistan is less affordable due to high transportation costs.

Furthermore, cotton growers are in desperate need of improved cotton seeds. Enacting a seed law requiring importing seeds that meet international seed performance standards is the most immediate way to improve seed supply.

The government should adopt a strategy to speed up adaptation research on introduced varieties, whether through commercial, government, or public–private collaborations so that farmers can get better seeds. There is also a need for the asset appraisal of the state cotton processing factories and the viability of privatizing them.

Since 2002, cotton has been cultivated on 33 thousand hector land with 36.3 thousand tons of production. The local cotton production has had a slightly increasing trend since 2016 [8]. Most textile firms use cotton as raw material as there is a lack of artificial textile raw materials such as nylon, polypropylene, and acrylic inside the country. Textile manufacturers mostly use cotton and wool from the domestic market, and synthetic fibers are imported from foreign countries. As the nation lacks cotton processing plants, a large proportion of cotton is exported for extra profit. Domestic production costs would be cut significantly if this cotton is not exported and manufactured locally since the same raw material would be obtained at a low price.

Although the natural textile raw materials, i.e., cotton, silk, and cashmere production, have a high production scope, artificial textile manufactured goods such as nylon, polypropylene, and acrylic are not available inside the country. As a result, textile mills often use homegrown cotton, while synthetic fabrics are imported from abroad.

3.6 Inefficiency in Textile Production

Productivity is poor, and manufacturing costs are high in the country's textile industry. As a result, Afghanistan's the textile and garment industry needs new marketing and operations techniques and processes to increase performance and productivity.

The latest technology management and marketing are not the characteristics of an Afghan company. They use old working methods for textile processing and production. Three decades of war have destroyed the physical infrastructure of the textile and clothing sectors and isolated it from introducing modern technology, the latest manufacturing techniques, and total quality management. It is a weakness of the current textile industry actors that they do not have obtained such technologies. Therefore, established companies may face challenges if new investors implement modernization because the investor's per-unit production cost would be meager. Existing businesses could face difficulties as a result of such an implementation. Although such introduction may create difficulties for the existing firms, it is required for the industry to compete with the region's major textile-producing countries in the long run.

The textile industry in Afghanistan, like other industrial sectors, can be divided into various divisions, such as producing finished and unfinished products. The country barely has any textile or apparel companies that cater to high-end customers, such as leather jackets or suiting fabric. None of the supply chain categories listed above can completely satisfy the local market's demand for apparel and apparel items. Even there is a lack of cotton processing plants that separate the seed from the cotton. A cotton processing plant in Balkh province claims that most raw cotton is either degraded or lost due to a shortage of cotton processing units. In addition, since most high-value goods are purchased from abroad and marketed in the home market, it can be considered a viable investment opportunity. Moreover, massive opportunities for investment can be identified after exploring Afghanistan's garment and apparel industry.

The textile and clothing industry of Afghanistan lacks modern management & marketing techniques to improve efficiency and effectiveness. Thus, this results in low productivity and very high production cost.

3.7 *Economic Opportunities in Textile and Clothing*

Since 2002, the government of Afghanistan has promoted a business-friendly climate to build and maintain a thriving corporate investor as the country's economic engine, following the transformation from a centrally planned regime to a market economy, as illustrated in the Afghanistan National Development Strategy (ANDS) (2016). As a result, a vast amount of money has been spent revitalizing Afghanistan's crumbling economy; the construction, transportation, telecommunications, finance, and defense sectors have benefited the most from these domestic and foreign investments.

In comparison to the services industry, investment in businesses and production has been minimal. The garment and clothing industry has been widely underinvested, despite its immense promise and long-term profitability. International and domestic investment can place the textile and garment industry as one of the country's primary sources of economic growth, foreign capital generation, and job development with appropriate government engagement and funding.

However, currently, the Afghan government has an economic vision to provide a favorable environment for investors. This policy is drafted to present solutions to challenges and remove barriers to investment. With this policy, investors will get government support and attain huge investment opportunities, especially in the manufacturing sector. The offered privileges and facilities for the private sector are providing land for free or with minimum price and making tax-free machinery or raw material imports with no custom tariffs. Therefore, this can be seen as a golden opportunity for those who like to invest in Afghanistan.

The Afghan government has also offered incentives through building many industrial parks throughout the major cities. According to [2], the industrial parks' incentive is: they provide the land with already built basic infrastructure and facilities. The land price can be given in ten-year installments, provided that the business starts within fifteen months of occupying the land. Favorable tax rates and regulations will be provided to investors in this period. In addition, quick structure building permission and other facilities will be provided.

Ahmadzai [2] discusses that Afghanistan's textile value chain, i.e., cotton processing, manufacturing fabrics, and cloths, has a small number of existing working factories that cannot satisfy local demand. His Afghan textile industry's review revealed that its strengths are the cheap labor force, a rising economy, and domestic demand. The industry's deficiencies were described as a high reliance on cotton, lower productivity, increased raw material exports rather than finished products, and technical deficiency. Huge investment gaps in the supply chain, joint ventures with international firms, and the development of high-value products have all been described as business opportunities. However, imports of value-added textiles and the export of raw cotton posed risks to the industry.

The central statistics organization (CSO) of Afghanistan estimates a total population of 32 million in 2020. Out of which, about 12.7 million are dependents and 11.7 million people under the age of 15. The workforce in the country constitutes 14.3 million people. Afghanistan, on the other hand, is reportedly one of the low-wage nations. Therefore, the textile industry's strength may be its considerable workforce and low wage levels.

Despite the drawbacks and pitfalls of investing in a post-conflict country's textile and clothing industry, the businesses in this sector have high-profit margins. Moreover, each component of the supply chain has an enormous investment advantage, and investing in each component would yield high returns and generate economies of scale. It also aids in lowering the cost of production for firms in the textile supply chain.

4 Prospects of Textile Education in Afghanistan

Afghanistan should be considered more than a war and conflict area with much more to offer, especially in textiles. If education is institutionalized in the region, many sectors will start developing, and the country can become self-sufficient in the textile field in no time. Moreover, by making valuable use of its human and social capital, the country can move on from its war problems and provide a source of happiness and contentment for its citizens if proper educational opportunities are established.

There is no formal textile education in Afghanistan; however, there have been opportunities for Afghan textiles and their popularity worldwide that could be capitalized on if textile education is formally given to the students. The degree holders could directly use the textile degree here in fields like cloth coloring, textile designers depicting Afghanistan's cultural heritage, fashion designers using the culturally famous clothing pieces and artwork, and clothing technologists for helping the small and medium enterprises already working in the area. In addition, the textile industry's indirect use can be viewed in further teaching of the SMEs, designing the products, school teaching, and merchandising.

Various factors have been playing their part in restricting the garment industry from performing to its fullest, from which it can be deduced that textile education is not present in the region. A study indicated that lack of security and support from the government, stiff competition from the foreign market, absence of textile factories, and less control over the large volumes of imports of the garments are some of the facets of the Afghani textile market that hinder its way for becoming a developed economic industry within the country [10]. For this reason, textile education has not been yet developed, which advocates that besides having skills within the people of this land, it should have been seriously contemplated.

There have been collaborations between foreign projects and the Afghan government to start formal education for promoting textile within the country. It means that people are interested in making textile products if they are given an opportunity. Moreover, it would not be necessary for a foreign textile artist to come to Afghanistan to teach and support the people for promoting textile here if there was a region's own formal textile education for the residing people.

Lately, there have been offers from other countries for Afghanistan to support its textile industry as the foreign countries think that Afghan textile has the prospect and energy to mark its place within the international market. Its textile designs influenced by cultural inheritance are liked by foreign people and widely used for interior design. Uzbekistan had offered Afghanistan to improve its economy and help stability by investing in its textile industry [21]. With these prospects, it could be inferred that the Afghan textile industry is assumed to prosper, which can open doors for textile students and formal textile education in the region.

There has been an initiation of projects within Afghanistan that aim to support practical training of women and girls for textile production at their homes. In addition, the projects have generated funds for skills training centers that would help in product development through weaving, knitting, cloth and silk making, stitching and tailoring, and carpet making activities [4, p. 41].

Loans for graduates have already been provided to use this money to buy their required equipment to produce and sell textile products. In addition, the centers are present for cooperative management support for the students who want in this field. These aspects could generate numerous prospects for textile education, which has become necessary for improving the country's economy. Furthermore, many of the country's daunting political and financial problems could be solved if education, especially in textile, the country's strength, and cultural tradition, could be formalized and supported by the government.

Some vocational schools for home economics were found in Kabul [17, p. 4]. This provides evidence that some form of teaching for textile is already taking place in Afghanistan. An example from its neighboring country, Tajikistan, could be established in Afghanistan, where an intensive training course (ITC) from the Global Textiles and Clothing Program was launched with the cooperation of the Technological University of Tajikistan and Singapore's Textile and Fashion Industry training center [12].

Switzerland's government-funded it to promote textiles in developing countries and assist their employment and income generation. The same example could be used in Afghanistan's vocational schools, and foreign-funded textile training programs could be introduced within Afghanistan so that income generation could be improved and economic stability within this region could be observed.

It is not necessary to have high technology-induced sewing and stitching machines in Afghanistan to prepare modernized fashion clothes; however, first initial steps could be taken to formalize this type of education in vocational schools, and the government can take a step ahead if a large number of population is learning in these vocational schools so that foreign funding in this area could be maximized for its profit generation in the form of amplified textile education and positive results.

5 Conclusion

The restoration and development of Afghanistan's textile industry would generate significant income for the government and numerous job prospects. Foreign competitors, mainly Asian countries like China, are presently putting pressure on the nascent textile industry. Nevertheless, given the increased competition in the market, the Afghan textile industry needs to analyze its business models carefully and take necessary actions. This chapter examines Afghanistan's textile industry and discusses the prospects and relevance of textile education as a critical success factor in reviving the industry.

Substantial restrictions prevent the rapid expansion of the Afghan manufacturing sector. The technological expertise and skills are limited. However, by way of comparatively low labor costs, small-priced factory space, and potentially significant agricultural raw materials and resources, Afghanistan offers the textile industry favorable conditions. In addition, cotton, cashmere, silk, fur, skin and hides, in particular, are also available as home raw materials. The production promotion policies for these products can have significant impact on people's livelihood. Therefore, careful consideration of these crucial aspects by Afghan government and other stalk holders would considerably affect restoring the country's crippled textile industry.

References

1. Afghanistan National Development Strategy (ANDS) (2016). Government Islamic Republic of Afghanistan, Kabul. Available at: https://www.refworld.org/pdfid/5b28f4294.pdf. Accessed on Jan20, 2021
2. Ahmadzai, M. (2013). Investment opportunities in textile and clothing industry of Afghanistan. Afghanistan Investment Support Agency. Available online at: http://iccia.com/sites/default/files/library/files/Final%20Afghanistan%20Textile%20Industry%20Paper%20-%20MA.pdf. Accessed on Jan 15 2021
3. Asare IT (2012) Critical success factors for the revival of the textile sector in Ghana. Int J Bus Soc Sci 3(2)
4. Asian Development Bank (2003) A new start for Afghanistan's education sector. Retrieved from https://www.adb.org/sites/default/files/publication/27935/new-start-education-sector-afg.pdf
5. Azimy MW, Khan GD, Yoshida Y, Kawata K (2020) Measuring the impacts of saffron production promotion measures on farmers' policy acceptance probability: a randomized conjoint field experiment in Herat Province Afghanistan. Sustainability 12(10):4026
6. Becker J (2004) Afghanistan: the war and the media. United States and the others: global media images on "The War on Terror". Nordicom, Gothenburg
7. Bittlingmayer UH, Grundmeier A, Kobler R, Sahrai D, Sahrai F (2019) Education and development in Afghanistan: challenges and prospects. Transcript Verlag

8. Central Statistics Organization (CSO) (2018) Afghanistan living conditions survey 2016–17. CSO, Kabul, Afghanistan
9. Ehsan Bayat Afghan Wireless Business Development and Philanthropy in Afghanistan (2020) Everything you need to know about Afghanistan's textiles industry
10. Faqiri FM, Shams MS, Gul S (2019) Small and medium enterprises in the garment sector of Afghanistan: challenges and prospects. Kardan J Econ 2(4):70–82. https://kardan.edu.af/Research/CurrentIssue.aspx?j=KJEMS
11. Ibrahimi F, Rana KS, Choudhary AK, Dass A, Ehsan Q, Noorzai AU (2017) Effect of varieties and planting geometry on growth, yield and profitability of Kharif mungbean [Vigna radiata (L.) Wilezek] in southern Afghanistan. Ann Agric Res 38(2):185–193
12. International Trade Centre (2020, February 25) Fashion and sewing professionals in Tajikistan learn the latest techniques. Retrieved from https://www.intracen.org/news/Fashion-and-sewing-professionals-in-Tajikistan-learn-the-latest-techniques/
13. Moahid M, Khan GD, Yoshida Y, Joshi NP, Maharjan KL (2021) Agricultural credit and extension services: does their synergy augment farmers' economic outcomes? Sustainability 13(7):3758
14. National Statistics and Information Authority (2020) Government of Islamic Republic of Afghanistan, Kabul, Afghanistan
15. Noori AH, Choudhary AK, Dass A, Raihan O (2018) Effect of varying nitrogen levels on growth, development and yield of Bt-cotton (Gossypium hirsutum L.) in semi-arid region of Afghanistan. Ann Agric Res 39(4):390–397
16. Noorzai AU, Choudhary AK, Bana RS, Prasad R (2017) Growth behaviour, productivity and profitability of promising mungbean varieties in semi-arid region of Afghanistan. Ann Agric Res 38(1):78–86
17. Sherzad AR (2017) Education in Afghanistan: challenges and suggestions for improvement. Herat University. Report no. 45. ISSN 1619-3660. https://www.researchgate.net/publication/315810717_Education_in_Afghanistan_Challenges_and_Suggestions_for_Improvement
18. World Bank (2018) Afghanistan overview. Retrieved online on 2019/10/14, from https://www.worldbank.org/en/country/afghanistan/overview
19. World Bank (2014) Islamic Republic of Afghanistan agriculture sector review: revitalizing agriculture for economic growth, job creation and food security. World Bank, Washington, DC. AUS9779
20. World Bank (2017) Afghanistan development update. Kabul
21. Yeniseyev M (2019, April 11) Uzbekistan offers to help develop Afghanistan textile industry. Caravanserai. Retrieved from https://central.asia-news.com/en_GB/articles/cnmi_ca/features/2019/04/11/feature-01

Masaood Moahid Masaood Moahid is an assistant professor at the Agricultural Faculty of Nangarhar University, Nangarhar, Afghanistan, and has acquired a Ph.D. (in agricultural economics) from the Graduate School for International Development and Cooperation, Hiroshima University, Japan. His Ph.D. research was focused on agricultural finance and rural development. He has published six peer-reviewed academic articles to his credit. Mr. Moahid also holds MSc (Agri.) in agricultural economics from the University of Agricultural Sciences Bangalore, India.

Ghulam Dastgir Khan Ghulam Dastgir Khan is a Japan Society for the Promotion of Science's (JSPS) Research Fellow and a Ph.D. (Development Economics) Candidate under the Taoyaka Global Leadership Program at Hiroshima University, Japan. He has co-authored several peer-reviewed articles and collaborates with individual researchers and institutions on research projects related to agricultural and rural development, education, peace, waste management, public management, and trade and transportation. He holds two master's degrees; first, Master of Commerce from the University of Peshawar, Pakistan (2010), and second, MS in development policy from Graduate School of International Development and Cooperation, Hiroshima University, Japan (2016).

Namuna College of Fashion Technology: Pioneering in Fashion and Textile Education in Nepal

Rajendra Singh and Alpana Shrestha

Abstract Nepal is a beautiful country rich in culture and traditions and home to numerous religious and ethnic groups. The development of fashion design in Nepal is primarily due to the culture as the attire worn by the ethnic groups. So, the development of fashion is mainly categorized due to the weather and cultural influence. The modern-day fashion garment industry has started in the early 80's when due to the popularity of fashion scenes in India's Bollywood film industry. People started copying and dressing like the actors and actresses from the movies they watched. The development of the Nepalese fashion garment industry was due to the then Quota System provided by the Multi-Fiber Agreement, which led to the Agreement on Textiles and Clothing. At the same time, due to the basic needs of the modern-day growing middle-class family, Namuna Silai was established in the early 1970s by Mrs. Gyani Shova Tuladhar, the pioneer fashion/costume designer in Nepal. With the completion of near three decades of experience in the fashion sector, she Namuna Institute of Technology in 1997. In 2002, to develop the human resource required in the fashion textile sector, NCFT, in affiliation with Purbanchal University Faculty of Management, established the Bachelors of Fashion Design program as an intensive three-year course. In the year 2016, launched the Bachelor of Fashion Design and Management program. Nepal's only four-year course in fashion education is delivered in eight semesters under 144 credit hours. The college has made rapid progress in the region; the achievements are presented in this chapter.

Keywords Nepal · Namuna · Fashion college · Fashion design · Creativity · Fashion education · Talents · Exchange program

R. Singh (✉)
Textiles, Fashion Merchandising and Garment Technology, Namuna College of Fashion Technology, Kathmandu, Nepal

Fashion Technology Subject Committee, Purbanchal University, Biratnagar, Nepal

A. Shrestha
Namuna College of Fashion Technology, Kathmandu, Nepal

X. Yan et al. (eds.), *Textile and Fashion Education Internationalization*, Textile Science and Clothing Technology, https://doi.org/10.1007/978-981-16-8854-6_6

1 Brief History and Overview of Fashion and Textiles Evolution in Nepal

Nepal is a beautiful country rich in culture and traditions and home to numerous religious and ethnic groups [1]. While Western fashion has influenced many neighboring countries, Nepal has retained its traditional dressing styles, especially in rural areas. Conservative clothing styles are preferred more in Nepal, and many are discouraged from showing too much skin.

As diversity is prevalent throughout the country, many people follow their sense and style based on their traditions, culture, and region. For instance, people living in higher altitudes in the Himalayas wear long thick clothes to cover up their bodies against colder temperatures [2], whereas people from the Terai region, the southern plains, wear thin and soft clothes in a much warmer climate.

Nepal's many ethnic groups follow very ancient, distinct, and deeply rooted customs and traditions. One can, therefore, also find varied Nepalese traditional costumes based on ethnicity. For example, in the hilly regions, men traditionally wear Daura suruwal, and women wear Gunyu Cholo. The Newars of Kathmandu valley have Tapalan & Suruwal for men and Misalan & Haku Patasi for women. The Sherpas wear Bakhhu, and the Choudhary people wear a dhoti (see Fig. 1). These are just a few examples showing the variety of dresses and costumes worn by Nepalis according to their different cultures, traditions, and societies.

Although having rich traditions and cultures, the term neo-modern fashion has been widely adopted in Nepal lately. Up to the 1960's and 70's cloths mainly served the purpose of covering up the body. Even people belonging to the upper class had fewer clothing items in their wardrobes than they usually have today. There was a slight change in the fashion scene due to direct influence from the film industry, particularly India's Bollywood [3]. People started copying and dressing like the actors and actresses from the movies they watched. The upper and high-class people were mainly seen following such fashions and trends.

The boom of readymade fashion garments came into existence as a cluster of industries in the early decades of the 1970s and 1980s. This is because of the

Fig. 1 In Daura/Gunyu Cholo (left) Tapalan/Haku Patasi (Centre) and Bakhhu (Right)

United Nations Conference on Trade and Employment through the General Agreement on Tariffs and Trade in 1974 when the developed market economies imposed country-specific quotas from 1975 until 1995. The initiation of the quota system phase-out started in three phases until 2005. This had seen abrupt development of Nepalese readymade fashion garments, mainly exported to the USA's largest market. More countries with developed economies flourished in Nepal after its embrace of economic liberalization through the 1990s. Then Nepal was one of many developing South Asian countries that played a substantial role in the global readymade garment industry [4]. These mass-produced textiles had become a staple export from Nepal. The last few years have shifted towards a brighter future for garment industries by improving the product segment, workers' issues, and the environment. In the 1980s, the garment industry in Nepal boomed because of interest and funding from Indian exporters. Due to the product quota limits in India, exporters looked to Nepal to increase their production. This expanded production served to boost Nepal's economy and its reach on the global production scale. Thus, Nepal became a viable option for countries to produce and export various textiles. By the end of 2004, intense competition in the global garment market broke out after the World Trade Organization's Agreement on Textiles and Clothing expired. Nepal struggled to produce their competition out and subsequently saw a fall in revenue from garment exports. The Multi-Fiber Agreement, an international trade agreement that allowed duty-free access to the U.S. market for Nepal, also fell through in 2005 and further exacerbated the country's declining international revenue [5]. The international economic aftermath of the 9/11 incident in the USA also negatively affected the U.S market reliance on the garment industry in Nepal. The USA was the recipient of 87% of Nepal's readymade garments until 2002. In subsequent years, Europe, India, Canada, and even Australia have become the largest markets for Nepali garments, making up 90% of the country's exports.

2 Overview of Fashion Textile Education in Nepal

Namuna has been pioneering in the fashion sector of Nepal for nearly five decades, since the set up of Namuna Silai (one of the country's oldest fashion and tailoring firms) in Kathmandu in the early 1970s by Mrs. Gyani Shova Tuladhar, one of the nation's most prominent fashion/costume designers. With the completion of near three decades of experience in the fashion sector, Mrs. Tuladhar then established Namuna Institute of Technology (NIT) in 1997. NIT is probably Nepal's first professional school to offer career and employment-oriented short-duration vocational courses in various aspects of fashion and clothing construction. NIT has already catered for more than 2500 students in the country's vast fashion and design sector. Moreover, the fashion industry was proliferating all over the world, including Nepal. Thus after the successful running of NIT for about a decade, with an ever-increasing demand for a highly qualified and skilled workforce in the fashion industry, it was realized that there was also a need for higher education in the fashion textile field. Moreover, a more practical and hands-on education and the professional skill-oriented course were much sought after by the fashion textile sector in Nepal.

Therefore in 2002, to develop the human resource required in the fashion textile sector, Namuna College of Fashion Technology (NCFT), in affiliation with Purbanchal University Faculty of Management, established the Bachelor of Fashion Design program intensively three-year course. NCFT thus became the first college in Nepal to pioneer formal fashion design education in Nepal. Nearly 450 students have graduated from NCFT to date. The graduates produced by NCFT have been providing outstanding service to the sector both at home and abroad.

NCFT provides students with the most satisfactory academic experience, complemented by an exceptional curriculum, an outstanding faculty member team, superior teaching and learning methodology, excellent infrastructure, and mutual relationships with foreign design institutions. Highly professional faculties and experienced resource persons with degrees from renowned international universities offer education and training to students, emphasizing innovative thinking, creative skills, new fashion designs, and practical expertise. In addition, much importance is placed on developing students' critical analysis abilities that support original fashion ideas. All this is well recognized by stakeholders in the fashion designing and manufacturing sectors of Nepal.

In order to obtain a Bachelor of Fashion Design degree, students needed to complete 123 credit hours and a total of thirty-one courses (Fig. 2).

With a growing demand for qualified professionals in the international fashion textile sector, NCFT in 2016 and affiliation with Purbanchal University launched the Bachelors of Fashion Design and Management (BFDM) program. Nepal's only four-year course in fashion education is delivered in eight semesters under 144 credit hours. This course will be beneficial for those who intend to pursue a Masters's degree in fashion education (Fig. 3).

3 Modules Currently Taught Under Textile Education in the University

The Bachelor of Fashion Design and Management (BFDM) program at the Faculty of Management of Purbanchal University aims to develop students into competent fashion professionals for any fashion industry sector. It is based on the principle that graduates will spend a significant part of their working life in a constantly changing environment. Therefore, the students should have an opportunity to obtain a broad knowledge of the underlying concepts and reality-based skills essential in fashion designing and management.

Fig. 2 Namuna College of Fashion Technology Building

Fig. 3 Academic fashion show by graduating students in NCFT

Upon graduation, a student should be well equipped to function as a designer in a fashion business, fashion manufacturing facility, and related government services. In addition, the graduates should also have a variety of career opportunities in different sectors of fashion, including entrepreneurship, for creating much-needed jobs for others.

The BFDM program especially attempts to:

- Provide a thorough and advanced education in fashion designing that will enable students to be well equipped to face the fashion field's ever-changing national and global dynamics.
- Produce a highly skilled workforce in fashion technology to address the growing demands of the industry and assist in the nation's human resources development.
- Enable students to be well-rounded professionals in fashion technology fields to be entrepreneurs or employed in various organizations and industries.
- Make students eligible for Master's Degrees in fashion-related fields, Masters's in Business Administration, and other degree programs in Nepal and international universities.
- Expose the students to the real industrial environment and current market trends.

4 Curricular Structure

The Faculty of Management recognizes the need for both breadth and depth in the total academic pattern. Therefore, the Bachelors of Fashion Design and Management (BFDM) degree curriculum comprises a fundamental, core, and advanced core components. The BFDM course is designed in three sectors: basic, advanced, and career/market-oriental. The core modules in the BFDM program are Sketching and Illustration, Pattern making, Clothing, Textile, Garment Technology, Computer, Embroideries, Management, Communication, Event Management, and Fashion Shows.

Students mainly undertake Textile courses from the 1st to fourth semesters of the BFDM program.

The Textile Material and Design course of the 1st semester provides knowledge about the textile raw materials—fibers, fiber source, and properties. The conversion of fiber into yarn, yarn spinning processes. It also provides the identification of the textile fibers. It is designed to provide basic knowledge on freehand sketching of motives essential for textile design. It provides comprehensive knowledge on color theory and helps develop the skill and master the selection of color. It further creates the skills to develop the theme with just one motif or motifs (composition) to support textile or apparel design which can be prepared as various print design, embroidery design, knit design, appliqué work, accessories, or any textile such as curtains, bedsheets, or table cloths. The main objective of this course is to enable the graduates to be able to:

- Understand the sources of textile fibers.
- Know the properties of textile fibers.
- Understand the uses of various kinds of textile fibers.
- Identify the textile fibers.
- Expertise in freehand sketching.
- Expertise in color selection and application.
- Get updated with Textile design and prints.
- Perform hand painting and able to transform the design.
- Gain knowledge and skill on Texture and Patterns.

  The Textile Design and Fabric Manufacture the 2nd semester provides knowledge about the various loom, weaving processes, and types of weave designs. It also provides knowledge about knitting and various types of knits. The main course objectives are that after the completion, the graduating students will be able to:

- Identify the loom parts and loom types.
- Understand the weaving process involved in producing woven fabrics.
- Understand the knitting process involved in producing knitted fabrics.
- Design basic weaves.
- Design textile prints.
- Get updated with textile design and prints.
- Perform hand painting and be able to transform the design
- Gain knowledge and skill on texture and patterns
- Expertise in designing textile knits, weaves, and crochet items

  The Textile Chemistry course of the 3rd semester provides knowledge about dyeing and printing textile materials with different dyes. It provides knowledge about dyeing machines used in textile industries. The course provides knowledge about printing techniques and the finishing of the textiles as per their use.

  The course is designed to provide basic knowledge on Freehand sketching of motives essential for Textile Design. In addition, it provides comprehensive knowledge. It further creates the skills to develop the theme with just one motif or motifs (composition) to support textile or apparel design. The main course objectives are that after the completion, the graduating students will be able to:

- Gain an overview of processes involved in dyeing, printing, and finishing textile materials.
- Familiarize students with chemical processing and its role in fabric manufacturing.
- Expertise in color selection and application.
- Design the Textile Prints.
- Perform hand painting and able to transform the design.

The Fabric Design and Analysis for the 4th semester provides a theoretical and practical introduction to how the fabric is manufactured in textile industries, focusing on weaving. The students are taught about different weaves of fabrics used for fashion design products and accessories. They will also study Managing People in Textile

Industries, Costing of Textile Materials and briefly learn about Production Planning for Textile Materials. The main objective of this course is to enable the graduates to be able to:

- Understand and differentiate between different types of woven fabrics.
- Managing People in Textile Industries.
- Costing of Textile Materials.
- Brief Production Planning for Textiles Materials.

Likewise, students are introduced to Fashion Merchandising. The course is designed to provide them with theoretical and practical knowledge and skills related to fashion merchandising and the business of fashion. The main course objectives are that after the graduating students will be able to:

- Carry out the merchandising of fashion garments/accessories.
- Carry out production planning in garment industries.
- Keep updated records and calculation of fashion garments/accessories financial status of the company.

5 Opportunities and Challenges of Textile Education in Nepal

In Nepal, the textile trade represents over US$ 675 Million (source: www.tepc.gov.np) worth of goods every year, out of which fashion goods export trade represents over 30% of the total export trade, amounting to over US$ 270 Million annually. In addition, the total fashion garments import value was over US$ 208 Million in the fiscal year 2019/20 alone. The overall export amount and their share from Nepal to different countries are illustrated in Fig. 4. Therefore, there is a massive opportunity in the fashion sector for trade, investment, and employment.

Fig. 4 Representation of exports from Nepal to different countries. **a** Export values in billions (NPR) to different countries. **b** Share of Export from Nepal to different countries [6]

A growing adult workforce is estimated to enter the labor market by over 500,000 every year. However, due to high competition in the job market, the workforce coming into the fashion industry often consists of cheap labor. It is essential to point out here that a good %age of the uneducated population can be catered into the fashion textile sector with concise on-the-job training.

It is apparent that the domestic industry in Nepal also needs fashion textile education to fulfill the growing demands of a wider group of fashion followers and trendsetters in the fashion garments sector. This has prompted the opening up of a growing number of small training institutes within the country. The education in such small local fashion institutes, schools, and colleges primarily focused on training people with little or no formal education, especially women. They usually cater to school dropouts, housewives, and part-time professionals with no age and education bar.

While there are enormous opportunities for decent jobs for fashion textile graduates capable of undertaking the entire fashion manufacturing and business functions into their responsibility, there is only a minimal number of suitable university graduates.

With an ever-increasing influence of western culture on Nepalese lifestyles, the local fashion industry is fast expanding. Job prospects for the graduates of fashion design are therefore promising. The global fashion industry and rapidly developing technology have also had a significant impact on the Nepalese people. The youth, in particular, have their unique style of dressing and are setting interesting fashion trends by following those from Western cultures. The lifestyle changes that are now apparent among Nepalese people have made the Nepali fashion world more sophisticated and alluring.

Therefore, fashion textile education offers the growing young population excellent opportunities to develop Nepal's significant textile fashion sector. However, as already mentioned, only one Purbanchal University affiliated NCFT is conducting a four-year Bachelor of Fashion Design and Management degree course in Nepal. So more colleges could be catered into the fashion and textile field to develop this vital sector further. It should also be acknowledged that there are limited lecturers and faculties in Nepal for higher education in fashion textile.

6 Fashion Centre

When traditional retail stores are floundering, department stores, superstores, supermarkets, and malls have emerged as the shoppers preferred choice to go shopping in Nepal; refer to lobby view in Labim Mall in Fig. 5. Business in these purchase points is booming thanks to fashion garments being the top products on display for sales. Many such stores have come into operation in various urban districts in Nepal, where customer and consumer numbers are increasing significantly. Supermarkets, malls, and other superstores now offer customers a much more comprehensive range of affordable clothing and accessories to choose from. One of the biggest ongoing

Fig. 5 Lobby view in Labim Mall

development trends in the country is the increasing accessibility to general fashion merchandise that is stylish yet affordable to middle-class families.

More people can be seen dressed up in finer clothing than in the past. However, it does not stop there as the trend is only increasing. Therefore, it will be even more interesting to see how fashion and styling develop in the future around the world and Nepal.

7 Collaborations in Fashion Field Development

Since its establishment, NCFT, with affiliation to Purbanchal University, has collaborated with international institutions to organize various programs and activities that helped further enhance efforts towards developing continuing learning methods and sharing experiences in fashion education and practices. The institutions with which NCFT has collaborated include:

- Raffle's Design Institute, Singapore.
- Raffles College of Design and Commerce, Sydney, Australia.
- Griffith College, Dublin, Ireland.
- Nepal Pashmina Industries Association, Nepal.

- Garment Association of Nepal, Nepal.
- Pidilite Industries, India.
- The British Council, Nepal.
- International Trade Center (ITC), Switzerland.

8 The Government of Nepal, Ministry of Industry, Commerce and Supplies: Technical Committee for Determination of Fabric Consumption for Readymade Garments

The Government of Nepal has duty draw-back facilities for imported fabric for use in the country's export-oriented fashion readymade garment industries. In order to monitor the consumption of such materials so that imported fabrics are not used to produce readymade garments for local sales, the Government of Nepal, under the Ministry of Industry, Commerce and Supplies, has set up a technical committee that determines the fabric consumption according to the size and design of the readymade garments meant for export to international markets. As a pioneer in designing fashion garments, NCFT represents the committee to give expert technical advice for determining the consumption of fabrics in export-oriented readymade garments.

9 Pashmina Enhancement and Trade Support Project, EIF, the Project of the Government of Nepal, Implemented by International Trade Centre, Geneva (2014–2017)

The project was aimed at responding to the trade-related development priorities of Nepal as identified in the NTIS. This goal also aimed to build sustainable local capacities and leverage additional resources from development partners. The project targeted reducing poverty in Nepal by generating income and employment opportunities while contributing to export growth and market diversification.

The project strengthened the competitiveness of the Chyangra Pashmina® (C.P.) sector through interventions along its value-chain (farmers, processors, exporters), targeted marketing in selected priority markets. In addition, the complementary strengthening of the C.P. sector's institutional support infrastructure also improved access to critical markets and trade-related information for producers, processors, and exporters.

The project achieved long-term sustainable impact by strengthening national capacities and Enhanced Integrated Framework (EIF) ownership throughout the implementation phase. In addition, the project also contributed towards achieving the following key results:

- Export quantity and value of Chyangra Pashmina® (C.P.) exporting firms increased.
- Nepal Pashmina Industries Association (NPIA) enabled it to provide priority services to its members.
- The number of business contracts along the C.P. value chain increased due to sensitization and facilitation of private sector investments.
- C.P. brand perception improved in selected target export markets.
- Enforcement of C.P. trademark in international markets strengthened.
- Viability was established for new modalities for increased local value addition through "de-hairing" or "de-greasing" cashmere wool.

10 Way Forward

There is a growing community of like-minded readymade fashion garment manufacturers, brands, and designers who share a true passion for ethical fashion and sustainable practices in Nepal. In addition, a growing number of young people are interested in fashion and seek knowledge to further Although there remains substantial work to be done to transform the garment industry into a thriving industry and an equally ethical one, Nepal is making initial steps towards successfully achieving both. This change will provide garment industry employees a better quality of life while ensuring that they and their families receive fair treatment at the same time.

Nepal is included in the Least Developed Countries' list by the United Nations Organization. Nepal is therefore granted primarily preferential duty-free benefits by most countries with developed economies. Despite this, Nepal could not maintain the necessary growth in exports of its products, including readymade fashion garments. As a result, there was a decline of over 23% in the exports of readymade fashion garments in the fiscal year 2019/20 from a value exported US$ 40.8 Million (Rs. 4.87 Billion) compared to a value of US$ 60 Million (Rs. 6.34 Billion) exported in the fiscal year 2018/19.

Nepal has, in recent years, adopted more intricate readymade garment designs using a wider range of materials, colors, styles, fitness, and better product quality for exporting. The importing countries in the developed market economies are willing to pay extra for the fashion products produced in Nepal's low labor wage rate. Continuing to focus on producing quality over quantity will positively impact Nepal's garment industry.

Current fast fashion practices can also be prioritized by creating as many items as possible within a set amount of time while shifting to more ethical work forms. This will ensure that the quality of garments continues to improve, raising the value of each item, bringing more foreign currency into the Nepali economy. The exact reach and impact that the garment industry has had on Nepalese poverty remain unclear, but the future looks bright. The Nepalese government reports that employment data within the garment industry is not readily available, but at its peak, the garment industry employed 12% of the overall labor pool in the Nepalese manufacturing sector. The modern garment industry, regulated with a minimum wage, can help lift Nepalese workers above the country's poverty line.

Nepalese fashion garment industries and companies have been working together to promote better transparency and industry practices for making them more sustainable. However, the fashion education sector in Nepal is very young and needs further collaborations to develop the multi-dimensional changes taking place in it. Nepalese fashion education should also incorporate collaborations with leading international institutions and organizations with the latest hands-on experience.

At the moment, for a start, the experiences of Donghua University in China could share with us the faculty exchange programs on the teaching and students exchange programs to get knowledge of the best practices done in the development of fashion education and profession. Likewise, the projects aimed at the Belt and Road Initiatives of China can be another greater opportunity for Nepal in fashion and education.

11 The Popular Graduates: In Fashion

Several graduates from NCFT are practicing designing independently soon after completing their graduation. Some famous graduates of NCFT in the fashion industry of Nepal are presented in Table 1. Apart from running their boutiques, some individuals with this degree are working as: Creative Fashion Designers, Fashion Merchandisers, Garment Technologists, Fashion Bloggers, Fashion Vloggers, Fashion Journalist, Costume Designers, Fashion Stylist, Fashion Academician, Trend Forecaster, Fashion Illustrators, Pattern Designer, Fashion Accessories Designer, Fashion Trend Maker, Embroidery Designers, Fashion Industrialist, Fashion Coordinator, Apparel Production Manager, Fabric Buyer, Fabric Quality Control Manager, Cutting Masters, Sales Manager in this industry and many more. The tremendous success of NCFT graduates as entrepreneurs and career professionals in fashion has contributed to consolidating further NCFT's status as the veritable pioneer and the pre-eminent name in fashion education in Nepal.

12 Conclusion

Nepal's textile trade represents over US$ 675 Million every year, out of which fashion goods export trade represents over 30% of the total export trade, amounting to over the U.S. $ 270 Million annually. Therefore, fashion textile education offers the growing young population excellent opportunities to develop Nepal's significant textile fashion sector. In order to develop fashion education in NCFT have been working to collaborate with international institutions to organize various programs and activities that helped further enhance efforts towards developing continuing learning methods and sharing experiences in fashion education and practices. Therefore, it would seek close collaboration with Donghua University in China, sharing the

Table 1 Famous graduates of NCFT in Fashion Industry of Nepal

Mr. Bishwo Gautam International Fashion Designer	**Mr. Yubraj Thapa** International Fashion Designer	**Mr. Manish Rai** International Fashion Designer	**Ms. Nilima Tamang** Fashion Designer and Stylist

Ms. Nisha Shrestha Fashion Designer and Stylist	**Ms. Pragya Chitrakar** Fashion Designer	**Ms. Pooja Shrestha** International Rug Designer	

faculty exchange programs on the teaching and students exchange programs to learn the best practices in fashion education and professional development. Likewise, the projects aimed at the Belt and Road Initiatives of China can be another significant opportunity for Nepal's fashion and education. We believe our already established number of graduates from NCFT who are practicing in the international arena would be the asset to showcase their talents and creativity in the design field more with this collaboration.

References

1. Gmińska-Nowak B, Ważny T (2020) Dendrochronological analysis of the ancient architecture of Kingdom of Lo, Upper Mustang, Nepal. Dendrochronologia 61:125701. https://doi.org/10.1016/j.dendro.2020.125701
2. Gautam B, Rijal HB, Shukuya M, Imagawa H (2019) A field investigation on the wintry thermal comfort and clothing adjustment of residents in traditional Nepalese houses. J Build

Eng 26:100886. https://doi.org/10.1016/j.jobe.2019.100886
3. Shresthova S (2010) Under India's big umbrella? Bollywood dance in Nepal. South Asian Popular Cult 8:309–323
4. Saha RC (2018) Regional cooperation in port development to bolster maritime logistics services in South Asia. Amer Soc Civil Engineers, New York, pp 142–155
5. Zanni D, Soria R, Smart P, Larsen J, Khanal S, Hate A (2005) The expiration of the multi-fiber arrangement: an analysis of the consequences for South Asia
6. Government of Nepal, M.o.I. (2021) Commerce and Supplies Trade and Export Promotion Centre. Data Visualization

Rajendra Singh has completed his postgraduation in Renewable Energy Engineering, Institute of Engineering, Pulchowk Campus, Tribhuvan University, Nepal, in 2004. Previously, he graduated with B.Sc in Textile Technology, Bangladesh University of Textiles, (Formerly College of Textile Technology, Dhaka University), Dhaka, Bangladesh, in 1996. He has experience for over 21 years in pioneering women entrepreneurship development through carpet weaving, dyeing and finishing, product developments in readymade garments, product development in pashmina weaving and dyeing, as well as other and other natural textile fiber products development by collaborating with local as well as international stakeholders to build the product development and export market linkages towards the national issue of managing the export promotion of these products, an effective trade policy program of Nepal. He has over 15 years of experience in teaching, currently Head of Department in NCFT, Kathmandu, Nepal in various subjects like Fashion Merchandising, Garment Technology, Fabric Design and Analysis, Textile Chemistry, Textile Materials and Designs, Textile Designs and Fabric Manufacturing. He has been involved in significant contributions in formulating national sector-specific export strategies of Nepal. He is actively involved as a trainer in export market development through the International Trade Centre, Geneva on market analysis tools and other export market development tools in the market intelligence market. He is also working as a consultant and actively working in core-teams and as navigators in formulating strategies for priority export products of Nepal by coordinating with the international development agencies like International Trade Centre, UNDP, and GIZ, on trade facilitation and product developments from Trade and Export Promotion Centre. He also mentors national stakeholders in inculcating them in the export procedure, business start-ups, market linkages, and product process development to achieve export development. He is also represented in different national steering committees and projects related to trade and investment in the Government of Nepal. Recently he has the concept of running Export House, Design Center, and Trade Facilitation Units in Nepal to get the trickle-down benefits up to the grassroot level SMEs in exports.

Alpana Shrestha is a graduate of Donghua University (formerly known as China Textile University). She first studied Textile Engineering and then Industrial Engineering at the same University in Shanghai. She has over 35 years of working experience in production, technology, machinery trade, publication, exhibitions, teaching, and consulting in textile-related fields. She has worked in Kathmandu and Hetauda in Nepal, Hong Kong, and Beijing and Shanghai in China. She is currently based in Kathmandu as Faculty in NCFT and serves as an independent consultant in the textile trade, technology, and media.

Evolution of Sri Lankan Textile Education from Ancient Times to the 21st Century

U. G. S. Wijayapala, A. A. P. Alwis, G. M. Ranathunga, and P. V. M. Karunaratne

Abstract Textile education in Sri Lanka has developed in its geopolitical historiography and is one of the oldest industries of Sri Lanka. The earliest handloom weaving textile industry was based on domestic weavers of Sri Lanka and practiced as an indigenous craft. Clothing has been a primary need of people and inhabitants of the country since ancient times. In ancient Sri Lanka, education was based on the system of inheritance; hence, it is considered the nation's wealth. Under this traditional educational system, the apprentice received specific training under a master for a period. Simple tools, local techniques, colors, intricate design motifs, and textures of textiles have been developed according to the country's existential contours of politico-economic and socio-cultural trends. The first textile corporation was established in 1956 with the government's patronage by collating domestic handloom weavers. Today the textile industry caters to domestic and global context needs. The unique selling point of the industry is indigenous tradition. A pedagogical network has been carried out to sustain the industry for years. This network consists of experts in art and tradition, technology, and marketing that bridge traditional knowledge and new knowledge. The textile pedagogy of Sri Lanka provides knowledge for the textile production process, which starts from yarn production to weaving of the material and caters to its assortment of uses. Education's meaning,

---

The original version of this chapter was revised: All authors' affiliations have been updated. The correction to this chapter is available at
https://doi.org/10.1007/978-981-16-8854-6_10

U. G. S. Wijayapala (✉) · G. M. Ranathunga · P. V. M. Karunaratne
Department of Textile and Apparel Engineering, Faculty of Engineering, University of Moratuwa, Katubedda, Sri Lanka
e-mail: samu@uom.lk

G. M. Ranathunga
e-mail: gayathrir@uom.lk

P. V. M. Karunaratne
e-mail: virajinik@uom.lk

A. A. P. Alwis
Department of Chemical and Process Engineering, Faculty of Engineering, University of Moratuwa, Katubedda, Sri Lanka

X. Yan et al. (eds.), *Textile and Fashion Education Internationalization*, Textile Science and Clothing Technology, https://doi.org/10.1007/978-981-16-8854-6_7

purpose, and beneficiaries were changed when European education standards were introduced and implemented in Ceylon. The early education was interwoven with occupation and the lifestyle of the stakeholder. The education system is so-called to the local context of the country. Even though the socio-cultural, economic, and political factors have changed, the context is the same as it was in the traditional society of Ceylon. Kandy was the last tentacle of the early administrative eras of Sri Lanka that lasted from the 14th Century to 1815 AD. The recorded history of Kandyan is rich with European documentation and local records. Retrieved artifacts of textiles are readily available in the Museum of Colombo and the museum of Kandy in Sri Lanka. The exhibits provided rich and credible evidence of textile education of the time. The apparel export industry in Sri Lanka started in the late 1970 caters mainly to the European market and is one of the world's top apparel-producing countries in terms of population, accounting for over half of the country's overall exports. Considering the high demand for textile designing, textile engineering, and related engineering disciplines, established related subject disciplines in technological Universities in Sri Lanka produce graduates to supply the demand of the textile industry. Some Universities offer textile designing degree courses that are based on humanities and multidisciplinary subject contexts. Today, certificate courses, diploma courses, and bachelor's degrees are offered by the Government of Sri Lanka and the private sector. Textile education has become in vogue in Sri Lanka because of the standardized tertiary education offered by the government and non-government higher educational institutes. The present-day textile education in Sri Lanka can be divided into vocational education, undergraduate degree education, and postgraduate degree education. The textile education of Sri Lanka is driven by two major pathways, which are the traditional domestic handloom textile industry and export-oriented apparel industry-related education. Education provides the nation with knowledge, skills, and techniques and expands the capabilities relatively to time necessities. The education imparted many advancements in major two streams sustainable survival. The chapter discusses the journey of textile education in Sri Lanka.

Keywords Textile · Education · Handloom textile industry · Local context · Export-oriented apparel industry · Knowledge · Skills · Technology

1 Introduction

Learning has had a long history in Sri Lanka. Religious instruction was given to novices in Buddhist temples well before the Christian era, and interested laymen were able to learn reading, writing, and religion at temples from about the same time. De Silva and De Silva ([39]:1) state that the traditional education system fosters respect for the 'educated' man. In theory, the system of temple education was open to all, though, in practice, the rigidities of the caste system restricted administrative positions to specific groups and thus might well have reduced the incentive to learn basic literacy skills, among others. Technical skills such as knowledge of engineering art and crafts were transmitted from generation to generation either by instruction from father to son or by craft apprenticeships. The traditional education system had been subjected to change due to the Western colonial occupation's conquest of lower lands.

Knowledge transmission and diffusion across generations are fundamental processes in all human societies. Individuals' learning styles, what they learn, and whom they learn from become part of their identity. At the same time, cultural expressions of traditional forms of indigenous knowledge have evolved in response to the preservation of local societal traditions. For historians interested in how crafts learning occurred in the past, theoretical perspectives on learning and historical documents describing learning provide background. This chapter gives a detailed description of the textile education of Sri Lanka from ancient times to the 21st Century the textile education is observed. The first half of the writing introduces how textile education was evolved with traditional customs, norms, and rules, which still exist and are charismatic for its survival. A change occurred in the early 19th Century relative to European settlement in the country. The western way of education has penetrated the country. The westerners interpreted the Existing education system. Industrial schools established by the British Sinhala Nationalist and the Education Department of Sri Lanka opened a new path of textile education. Policy orientation in education or policy formulation and planning paved the way to educational reforms in the Republic of Sri Lanka. Pre-vocational studies were introduced. The government institutions promoted technical education. In 1960 government departments were established to promote the cottage industry. These institutes deliver teaching, commerce, and craft-level training courses.

The 21st Century marked the evolution of textile education, including the Sri Lanka export-oriented garment industry and textile sector. As a result of these upheavals, government and non-government textile education started Universities and institutes. The final part of the chapter focuses on discussing issues and possible solutions related to existing textile education.

2 Sri Lanka's Textile History

2.1 Traditional Textile Industry, Learning Pattern in the Early Administrative Eras of Sri Lanka

Sri Lankan culture was developed based on the foundation of Buddhism. The ancient vocations and the learning system also developed on this same grounds [45]. According to the legends, Buddhism indicates the textile industry and its learning patterns [13]. The role of traditional vernacular education in the country can be best understood in terms of the total education system, which functioned in the traditional village society. The traditional community in which all the parts fit into an organic whole in which an appropriate value system supported the mode of production. Buddhism provided the basis for the value system and ethos of society. The agrarian economy was subsistence with the ancillary crafts and occupations supporting food production for the community. Society was organized based on occupational castes [40]. Each occupational caste provided services to the community [6]. Both females and males have gradually understood the local 'textile community of craft practice' by beginning to learn textile techniques as 'novice' learners from the 'master' members of their community. Some references indicate that the individuals are intended for the occupations [8, 14], thus continuing the hereditary production system by their birth occupations themselves [6]. Apprenticeship is common across cultures, but it is far from universal.

Furthermore, a formal apprenticeship may be limited to one or two crafts or skill sets in each society. Within a system, the individual member is given skill and introduction to occupation and craft, the education into a society's value system. Learning and work were linked closely together. The development of social ties is crucial to this learning process.

According to literary sources, individuals are predestined for their respective occupations at birth, perpetuating the hereditary pattern of output. The following are some of the families that are destined for those occupations. In one of her previous lives, Mahā Pajāpati Gotamī was born into the family of the chief, textile worker (*pesakārajeṭṭhakassageha*) of the textile workers' village (*pesakāragāma*), which was near the town of Bārāṇasi. It is also mentioned that in a previous birth of Ven. MahāKappina was born as the chief textile worker (*jeṭṭhapesakāra*) in the textile workers' village (*pesakāragāma*) near the town of Bārāṇasi. He has seen and used gold for a very long time after having been born in the families of goldsmiths (*suvaṇṇakārageha*) for over five hundred births. As a result, the subject of impurity (*asubhakammaṭṭhāna*) meditation is not appropriate for him [20].

A ranking system among occupations could be defined based on literary evidence. Some jobs have been regarded as inferior, and others have been regarded as superior. According to Sumangalavilasiniya crafts (sippa) are divided into inferior (hīna) and superior (ukkaṭha), and textile works belong to (pesakārasippa) inferior crafts [20]. Learning is thought to be a consequence of doing something (learning by doing). Visuddhimagga[1] stated institutions for learning and training such as the places of learning work (kammāyatana), places of learning crafts (sippāyatana), and places of learning sciences (vijjāṭṭhāna). As per the lexicographical meanings of *kammāyatana* and *sippāyatana* could be the locations of education and training.

Education's efficacy will serve as a social mobility agent. At the same time, the presence of educational and training institutions can call into question the occupations' hereditary nature. This is particularly true as professions requiring intellectual abilities rise to the top rankings [20].

[1] Visuddhimagga is the Buddhist practice's 'great treatise'. It is a manual that condenses and organizes the Buddhist path's perception and interpretation in the fifth century.

2.2 Historical Roots of Origin of the Sri Lankans Linked with Indigenous Weavers

The recorded history of the island and its people start with the Aryan colonization marked by the advent of Vijaya and his seven hundred followers from North India in the 6th Century BC after establishing Indian-influenced administration in the country. The chronicle *Mahavamsa* states that when Prince Vijaya arrived in Tambapanni (between 250–210 BC), *Kuveni, a* Princess who belonged to the *yaksha* clan, had been spinning cotton. It was the first written fact about textile weaving in Sri Lanka. (Mahavamsa 1959, 07:11) *Mahavamsa*[2] citing by ([10]: 28), mentions that as per the request by King Vijaya, the King of Madura sent him one thousand families of eighteen different "*Senis*" or Guilds, including weavers [7, 10]. Furthermore, there is a saying that the wedding ceremony of Prince Vijaya and Princess Kuveni was decorated with yards and yards of clothes as carpets (*Pāwada*) and hoods (*Udu viyan).* The *Mahavansa* states that the emperor Dharmashoka (268 to 232 BCE) in Bharatha once sent two fabrics as a gift to King Devanampiya Tissa (250–210 BC) when the clothes got dirty, that special fabric was cleaned by burning it in the fire. *Sahassawattupakarana* citing by [10], says that King Dutugemunu (161–137 BC), while ruling Magama, bought 100 Chinese fabrics, 100 white clothes, 100 fabrics with prints of wolves and animals, 100 fabrics with floral prints from one of his friends who was engaged in trading. *Sihalawattupakaranaya* citing by [10], stated a story about how King Saddhatissa's (137–119 BC) wife's brother offered rough clothes (*poththi/dala pili saluwak*) to a priest in need.

2.3 Historical Records Provide Historical Information of Textile Weaving, Weavers, and Textile Imports

In "*Kapu Upatha*[3]" a collection of folk poetry said that cotton was first brought to Sri Lanka by the great King Manu Aka Mahamandathu, the first ruler who started the dynasty of Kings [10]. *Rajatharangani*[4], citing [10], stated that royal princesses of Cashmere had imported valuable textiles from Sri Lanka. The literary work *Poojawaliya*[5] [47] recorded the immensity of the textile industry in Sri Lanka. There is information related to textile weaving on stone inscriptions in North Central

[2] The ancient Pali chronicle of the Sinhalese people.

[3] Kapu upatha (birth of cotton) also known as Kapuyakkariya, which is a part of 12 events that took place ***Rata** yakuma,* also known as *Riddi yagaya,* a traditional healing ceremony especially performed for barren ladies and for pregnant mothers to bless. Kapu upatha which depicts the steps of making a cotton cloth. At this time, the songs related to the origin of cotton will be sung.

[4] *Rajatarangini* (*Rājataraṃgiṇī*, "The River of Kings") is a metrical legendary and historical chronicle of the north-western Indian subcontinent, particularly the Kings of Kashmir. It was written in Sanskrit by Kashmiri historian Kalhana in the twelfth century CE. The work consists of 7826 verses.

[5] Pūjāvaliya -Pāli Commentary (thirteenth century CE).

province (Rathmal wewa). The presence of manufacturing-related seven types of occupations is evident in the Brāhmi inscriptions, the earliest existing records on the island. Among them, a weaver (*tantavaya or tantavaya*) was of the noted occupations [32]. *Cūlavaṃsa* (12th Century CE) refers to nine types of occupations and weavers identified as pesakāra [12]. Vitharana [48] also summarizes the 16 occupations as appearing in the Sinhala texts during this time, thread spinners (nūlkāra) and weavers (pehera) were included.

King Mahadathika Mahanaga (7–9 AD), the grandson of King Devanampiyatissa (250–210 BC), had woven *Uththaraga* (single-layered robes for priests) and offered them to Buddhist priests along with rice porridge. In one inscription of the first century AD, a weaver is referred to at *Kaduruvava*, a tank named *pehekaravavi* (weavers' tank). Galkandegama inscription mentions a village named *pehekara* that denotes the village of the weavers [10]. Ellawala [10], citing *Sumangalavilasini*[6] and *Abhidharma Pradeepika*[7] stated some of the other early literary works in which cotton (*kapu*) utilization is described. When King Dutugemunu (161–137 BC) was sick and moribund, his brother Saddhatissa ordered to weave yards of white clothes to cover the great *RuwanweliSeya* (a Buddhist monument) which had been under construction. King Gajabahu I (114–36 BC) of the Anuradhapura era has invaded South India and has brought back 10,000 Sri Lankan prisoners along with technicians, including weavers. The weavers were called *Salagamayo*. When Ven. Mahinda Thero first brought Buddhism to Sri Lanka during the reign of King Devanampiyatissa (259–210 BC); along with Buddhism came technicians of 18 castes based on industries, including weaving.

Mahawansa [14] stated that valuable clothes were received from Burma and Thailand in the 11th Century. Eulogistic verses sung in the King's court (*Raja sabha wenuma*) during the Kotte period (1371–1597 AD) records in the poetical work known as *Hamsa Sandeshaya* (Messenger poetical work, which the messenger is a swan) says that the King in the kingdom of Kotte (1371–1597 AD) wore silk attire with work of gems and pearls on it. With the ancients' beliefs, there are stories about clothes given by gods. They are mentioned in ancient literary works such as *Sahassawaththupakarana*. The minister's wife named "*Sangha*" once offered her valuable cloak to a monk and wore a rough piece of cloth instead. So, she was rewarded by God. A woman named "*Kundisangha*" offered her one and only cloth to a monk. Gods were impressed by her act and offered a fine, gracious cloth to her. These migrants settled in the country and contributed to the development of industries in Sri Lanka.

During the kingdom of Kotte ((1371–1597 AD), the handloom weaving industry was well developed when the Hollanders arrived. There were records found in journals of foreign travelers who have visited places like Batticaloa, Chilaw, Ambalangoda, Hambantota, and Galle. Puttalam, Chilaw, Ambalangoda, and Mannar were

[6] The Sumangala-Vilasini, Buddhaghosha's commentary on the Digha Nikaya.

[7] Abhidharma (Sanskrit) or Abhidhamma (Pali) are ancient (third century BCE and later) Buddhist texts which contain detailed scholastic presentations of doctrinal material appearing in the Buddhist sutras.

famous for weaving rough clothes, while soft fabric using thin yarn was woven in Jaffna [4, 34]. Also, the journals stated that cotton cultivated in Sri Lanka was even more eminent and quality than Brazilian cotton. King Prakramabahu II (1236–70 AD)—patronized the ceremonies of offering robes to Buddhist monks (*Katina civara)* during his reign more than once. Chapter LXXXV of *Mahavansa* says about this offering. The King Vijayabahu III (1232–36 AD) sent a muslin cloth by an envoy Pathi Meera Lebbe to South India with a letter and got down eight textile manufacturers to develop the industry in the Dambadeniya period (1232–84 AD) [7, 38]. The *Dambadeni-Asna,* or the story of Dambadeniya, is a short history of the reign of Parakramabahu II (1302–26 AD) mentions 111 types of textiles that had been used in the era.

Ancient literary works such as *Dampiya Atuwa Getapadaya*[8] and *Vishuddi Marga* texts made planer; homespun clothes mentioned many types of textiles including *Thihiri pili, kapu pili* (cotton clothes), *Koseyya pata, Komu pili, Kasi salu* (clothes from Cashmere) were imported from abroad. (Somāra deshaya). During famines and droughts, both males and females wore pilothika or poththi, the rough cloth.

2.4 Textile Weaving Industry and Arts and Crafts Education During Kandyan Period

During the Kandyan period (14th Century AD—1815), Medieval Sinhalese (a religious and national art) was the most popular and honorable of all occupations when the textile weaving industry was developed. Farming was an integral part of the civil service, and the craftsman was rewarded with state land grants called *rajakariya* [6]. Coomaraswamy [6] conclusion about art and crafts education imparted by tradition was the national system during the Kandyan Kingdom, where every civil person's education and occupation were interlinked. During this period, two groups of weavers called Salagamayo made fine gold-woven clothes and *Beravayo*, homespun clothes [6]. *Salagamayo* had been continuing in weaving since King Vijayabahu III (1232–36 AD) of the Dambadeniya period Commaraswamy [6] says that "the weaving of homespun cotton cloths was once universal in the Kandyan province and still continues weaving". Weaving has been carried out from generation to generation by certain families as a domestic industry. Both males and females are competent in this industry and doing it at their homes. They were well educated on the craft, from cotton grown to the finish of the product.

Further, Commaraswamy [6] says that the first and most thorough part of the Sinhalese craftsman's education consists of a particular craft's close and lifetime study. The craftsman carried on practice and spent time copying and repeating patterns until he gained such command in the crafting. Commaraswamy [5] says

[8] Is a commentary work written to explain the obscure instances of the Buddha's teachings a mainstream Buddhist work as well as connotation to the obscure instances found in the original work.

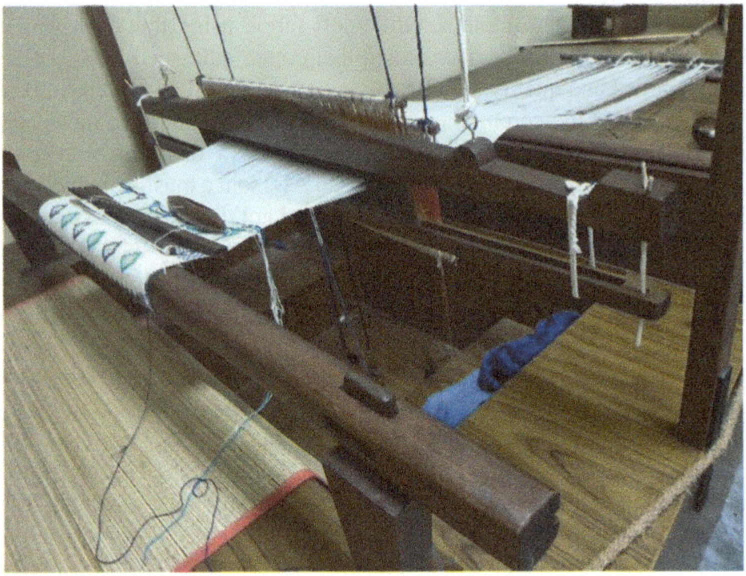

Fig. 1 A replica of old pit loom[9] at technological museum at Polonnaruwa

that the manner of working in looms is identical in all essentials with the looms and methods in use all over Asia. The loom was set up in an open shed of the weavers' house. Therefore, the weaving of the Kandyan Kingdom has been continued from generation to generation by leaps and bounds.

A range of products that belong to the Kandyan era is on display at museum collections of Sri Lanka. Those are bedspreads, handkerchiefs, *tuppotti* (body cloth for men), *hela* (saree for female), *diya-kacci* (loincloth for King), kerchiefs, shawls, pillowcases, belts, mats, carpets are some of them. Blue and red stripes on white cloth are common [46]. Varieties of colors are rare. Textile colors were red, blue, on white. Many designs and motifs are so-called to the Kandyan Kingdom and have continued in the current society giving retrospective value. Textiles and handicrafts resemble four categories of traditional Kandyan motifs. Since the days of the Kandyan era, villages called Dumbara and Thalgune, situated in the hill country of Sri Lanka famous for handloom textile weaving, have continued the inherited tradition of indigenous textile (Figs. 1, 2, 3 and 4).

[9] Coomaraswamy [6] heddles are pushed by coconut shell pedals that hang down into a hole under the loom, through which the weaver's feet move up and down.

Fig. 2 Yakdessalage Gedara Dayawathie (mother) 7th generation of the ancestors of indigenous weavers of Thalagune, hill country, Sri Lanka

Fig. 3 Sachithra Piyumali Wickramasinghe (daughter) 8th generation of the ancestors of indigenous weavers of Thalagune, hill country, Sri Lanka

3 The Change in the System of Traditional Textile Education

By the time the Portuguese arrived in Sri Lanka (1505), the textile industry and the local weaving industry remained developed. The traditional industry was changed by the arrival of the Dutch in 1638. They signed a treaty alliance with King Rajasinghe II (1687–1707 AD) that the King should grant them the monopoly over trade on the entire island once they chased away the Portuguese in coastal areas. After receiving monopoly, the Dutch (1640–1796 AD) allowed importing various clothes to Sri Lanka. The Dutch also imposed taxes on people who were engaged in weaving. In 1808, the tax was only 5%. It was imposed by stamping the clothes they wove. However, government servants of the Dutch government were released from the

Fig. 4 An old bedspread produced by ancestors of Yakdessalage Gedara Dayawathie and offered to the village temple, Keulgama temple, Ududumbara. Now it is kept at Dayaeathie's home. Traditional Dumbara motifs: lion, bird, elephant, and geometrical designs are displayed on the white background, unique to Dumbara motifs. According to their records, this would be older than 100 years

tax and were privileged to engage in weaving at their houses. Local weavers were discouraged s by these new rules of the Dutch and abandoned the industry. However, a quantitative expansion of education occurred under British rule. The vernacular schools in the villages provided skills-based vocational training. The early stage of British rule in Sri Lanka (1796–1948) ushered in great change.

3.1 Establishment of the Industrial Schools: By the British Rule

From 1859 onwards, missionaries opened more than 100 industrial schools all over the island. They immediately offered disadvantaged youngsters training in agriculture and crafts. After acquiring the state, they gradually focused on training their customer base under the Department of Education and later under the Department of Industries for jobs in emerging public industries [44]. In 1893, in Colombo, the Department of Education opened a technical college based on the principles of technical colleges in

Britain, whose academic syllabus was developed to educate those employed at the upper levels of public works departments and upcoming industries [28]. In the years immediately before and after independence, i.e. during the most crucial era of recent education growth in the country, political proposals that progressively integrated professional education and practical oriented skills into school level education were discussed. Add optional subjects relevant to high school practice or create a lower secondary level screening system that will also direct the least capable students to be vocational and technical courses at the academic level [21]. The Ceylon Technical College was established in 1894 at Maradana. Besides providing skilled workers for various government departments. The Ceylon Technical College provided instruction in various branches of engineering science, commerce, and arts and crafts ([39]: A76).

3.2 Establishment of Industrial Schools: By "Sinhala" Nationalists

The last quarter of the 19th Century thus laid the foundation of the 'modern' educational system in Sri Lanka. The British model became the accepted norm ([39]: A22). A few technical education instructions were successfully established, such as practical training in commerce, handicrafts, and domestic science.

The first industrial training school in the country, 'Hewavitarne Industrial Centre', was founded by Anagarika Dharmapala and Walisingha Harischandra in 1906 ([51]: 64). The center offered weaving, pottery, sewing, and similar arts and crafts courses. Under this process, in 1906, the young scholar Mr. *Ukku Banda Dolapihilla* was awarded the "*Hewawitharana* Industrial Scholarship" to help uphold the textile industry by learning the subject. He spent five years in Japan learning textile engineering at the Imperial College of Technology, Tokyo. In 1911 he returned to Sri Lanka and saw that the available finances were inadequate for adopting the Japanese type of handloom. He went to India and worked under the weaving master of Banga Laximi spinning and weaving mills, Calcutta. He made the loom used today by using the knowledge gained by both countries as the upper half of the Japanese loom and the lower half of the Indian loom. In 1912 weaving was inaugurated at the premises of Maha Bodhi Society, Maligakanda, with limited sources having a single pair of looms. He started Rajagiriya Hewawitharana weaving school, admitting 15 male students and within six months admitting six girls. The improved fly shuttle loom was first introduced in 1913 in Hewavitarne weaving school, and this method would be able to increase the output 3–4 times.

The improved method spread to Jaffna and Batticaloa. With the assistance of Sir Anagarika Dharmapala, Dolapihilla promoted Sri Lankan people to buy locally produced clothes; their mission also aimed to boycott all British clothes and goods during the colonial period of Sri Lanka. Within two years, he successfully toiled to turn out 300 well-trained weavers. His students went out one by one to set up their looms in various parts of the country. The opposition from vested interests in the import trade increased as the industry seemed to develop all over the country. During the 1950s, the government gave the industry a helping hand. With the assistance of the government, Dolapihilla pioneered opening 28 weaving schools and 60 textile centers in 1941. In the same year, Dolapihilla was assigned as the textile technologist and specialist adviser in textile technology to the Department of Commerce and Industries. In 1956, the government set up the Textile Industries Cooperation and appointed him as the Director. Wickramasinghe ([51]: 64) suggests that the schools were modeled based on the industrial schools Dharmapala had visited in the United States, England, Italy, and Denmark. Technological change was an important feature of the school. The handloom was used as a modification of the English handloom. While modern methods were being tested, homespun thread using the *chakra* was also introduced at the school. In 1917 the state sponsorship was given to the handloom textile industry. Sharma ([37]: 5) says that Catholic and Protestant missionaries ran Western education schools that provided British educational tradition.

3.3 Establishment of Weaving Schools: By the Education Department of Sri Lanka

The Education Department trained two scholars abroad in the year 1920. The department started a government weaving school at Thalpawila and a school at Batticaloa in 1921. In 1936, during the Donoughmore experience in self-government, the Department of Commerce and Industries first increased handloom textiles. In 1942 there was an addition to a training institute at the Technical College, Maradana. There were 16 government weaving schools, and 12 assisted weaving schools were established. Demonstration centers were started based on the model of India based on five objectives.

- To demonstrate the learner's modern methods of processing,
- To provide facilities for deserving workers locally,
- To provide officers to visit homes or institutions locally and to give the necessary advice and instruction,
- To encourage and assist workers to buy their machinery such as looms on easy terms,
- To give whatever assistance is necessary to encourage and promote the textile industry ([51]: 65).

4 Policy Orientation in Education: Policy Formulation and Planning

One of the most significant catalysts in the change process was introducing a free education system in 1944 by the education minister, CWW Kannangara. The monopoly of a small English educated minority has been changed to the middle class in Ceylon. The national languages were adopted as the media of instruction in all schools. Opportunities were made available to the entire population of school-going age to proceed from the primary to the secondary and tertiary stages. Kannangara's mission was to introduce a universal free education system for Sri Lanka by identifying education as an investment in human resources. He did not accept the concept of Industrial Schools and promoted vocational training [30]. In 1931 he recommended a Central Crafts School. The Technical Schools operated under Kannangara. Weaving was added as a subject for poor people. He also believed that the learning by doing method is more relevant to the Sri Lankan context [42]. Introducing free education was driven by his concept called "development of the wealth of the hand, of the heart and the head". He introduced a special committee on education in 1935 and directed the Education Ordinance in 1939.

During this period Trade and Industrial Schools in the country. There were weaving schools operated at Horethuduwa and Hiramadagama. In 1936 he established the Central Weaving Institute. The Council of Industry Commission has recognized textile as a suitable industry to start in Sri Lanka. The textile was recognized as a domestic industry which includes yarn producing, textile weaving and producing T'Shirts (*mes bannian*) ([23]: 90). By 1944 there were 125 textile training centers conducted by the Ministry of Industries and Commerce ([23]: 91). The report has recognized caste-based art and craft as a barrier to its further development. The report has recognized that the caste-based craft industries could not change. The duration of the technical education is six months for the pupils of 16–18. It was recommended to establish training centers in traditional craft villages. In 1947 the Technical College curriculum included a new course, weaving.

4.1 Educational Reforms in the Republic of Sri Lanka: Introducing Pre-vocational Studies

Sri Lanka gained independence in 1948 from the British, which is called a Dominion status. In the country, educational expansion became an ambition of the native bourgeoisie before the country's independence. Not only did this elite government implement compulsory franchising, but it also laid the foundations of a democratic social structure in the 1930s, including the right to free education. In other words, by the time of political independence from Britain, this late colonial constellation represented a significant key moment that stimulated a rise in literacy rates of up to about 65%. The first significant reformation in educational administration came in 1966. This reformation had three major objectives. The first of these three was the elimination of duplication of work between the Ministry and the department under it, and the second was to ensure greater coordination at the center between education and technical training and education ([39]: A45). Fifteen departments were set up at

Anuradhapura, Bandarawela, Batticaloa, Chilaw, Colombo North, Colombo south, Galle, Jaffna, Kandy, Kaluthara, Kegalle, Kurunegala, Matara, Nuwara Eliya, and Rathnapura.

In 1971 Sri Lanka became a Republic and could be seen as a development in every aspect of education. The educational reforms were introduced at the beginning of 1972 by the Ministry of Education, and the new curriculum targeted the junior secondary stages (grade 6–9) in 1972. The new curriculum included ten subjects and included pre-vocational studies [42]. The introduction of pre-vocational studies 1 and 2 is the other major change in the curriculum. The Ministry of Education has used pre-vocational studies refers to one of the Secondary school instructional programs, and from 1960 to 1974, schools had a program for teaching handicrafts, including weaving, identified as light miscellaneous crafts. Most girls generally took this subject and taught only 500 schools island-wide for three periods per week. Introducing pre-vocational studies, 2 was the new entrant to the curriculum. This second part of vocational studies included cloth printing. The schools have been informed that they should register with the Ministry, the vocations they have selected for the 1st National Certificate Education exam to be held in December 1975 [42]. The sole idea of having established pre-vocational studies into the school curriculum was to bring the simple vocations into the school or take the school out into the vocations brought out forcefully. Educational authorities thought that pre-vocational studies give knowledge, skills, and attitudes to learners.

4.2 Establishment of Government Institutions to Promote Technical Education

Technical College at Maradana was transferred in 1960 to the institute of practical technology at Katubedda, Moratuwa, teaching commerce and craft-level training courses at the craft level. ([39]: A80). In 1966 the Institute at Katubedda was converted into a College of Technology Teaching, a five-year diploma course. They provided some courses at a professional level. These include the higher National Diploma in Textile (NDT). Apart from these organizations and the training schools run by the Sri Lanka Transport Board, the Industrial Development Board has a training institute that gives craft-level courses lasting six months to a year. The Department of Small Industries provided training on handlooms and power looms ([39]: A82).

In 1971, the National Apprentice Board (NAB) was established to coordinate enterprise-based apprenticeships. The objectives of the authority shall be as follows,

- The preparation, coordination, and provision of technical training,
- The establishment of standards relating to vocational training,
- Conducting checks, awarding certificates and other prizes,
- Conducting tests on national trade,
- Conducting technical training research and development,

- Holding competitions to encourage the growth of different skills,
- Have ties with institutions, share significant disciplines in Sri Lanka and abroad, and compare and justify credentials, diplomas, and degrees in associated courses.

The same institution was transformed as the National Apprentice and Industrial Training Authority in 1990 under the TVE Act. No. 20 of 1990.

In the wake of the economic resurgence after 1976, education in commerce came into favor. The Ministry of Higher Education, established in 1978, has taken over the area of Universities and Technical Colleges [39]. By 1977 pre-vocational subjects were renamed technical subjects and their content considerably modified. The range of these subjects was eleven. A plan for advancing tertiary and vocational education is provided by the Tertiary and Vocational Education Act [1]. The Commission's objective was to plan cooperation in managing the economy's human capital needs by offering vocational awards including qualifications and other academic distinctions, ensuring academic and training standards in institutions and agencies. Technical education in the 1980s is thus in a period of expansion. It is receiving an inflow of government and private resources as never before. Universities and other higher educational institutes were established under the Universities Act, No 16 of 1978.

4.3 Political Intervention: Education for Unemployment Youth; Establishment of the Education Systems and Early Expansion

The United National Party (UNP) continued a corresponding policy after 1948 because it had been instrumental in attracting voters and because it attracted voters. In the early stages of independence, the profits from commodity exports encouraged the further growth of the welfare state [42]. After 1956, when the leftist Sri Lanka Freedom Party (SLFP) came into power, progressivism became the core argument for education policy. Private and English secondary schools were banned, and more state funds were allocated to improve general education, offering the chance to ascend the social ladder to many young people [29]. Because of these combined measures, Sri Lanka is one of the first postcolonial nations to come close to achieving compulsory primary education.

In 1977 the United National Party (UNP) took over power. The new UNP government, funded largely by the World Bank and the United Nations Development Programme, complemented its market-oriented economic reforms after 1977 to expand realistic training opportunities (UNDP). For example, the government set up a training center for clothing operators to attract foreign investors to the new free trade zones, which was to be followed by establishing a training center for clothing technicians some years later [31].

Financial democratization and structural reform strategies from the late 1970s onwards were strongly affected by society's overall growth in Sri Lanka. With a left-wing government in 1977, this process started dramatically in Sri Lanka, bringing the UNP back in power, and was revitalized in the second half of the 1980s and after 1995 [16].

5 Establishment of Government Departments to Promote Cottage Industry

The department of small industries was implemented to promote and develop cottage industries/rural industries, and small industries invariably involved young persons of both sexes. In 1973 handloom registered trainees were 100,000 and the education level ranges from grade 6-GCE depending on the availability of suitable young persons in the rural areas. They received six months of training and are on the job. During 1950–1960, the industry expanded allowances of 75% per day per trainee paid by the government as an incentive for the unemployed rural youth to participate. In 1973 Sri Lanka produced only about 5–7% of raw cotton required by the textile mills, decentralized power looms, and the handloom industry to meet the country's demand for textiles. Textile education was supported by workshops arranged by the Department [30] (Chart 1).

In 1964, the department organized power loom workshops in rural areas, resulting in the production of 25 power looms (Chart 2).

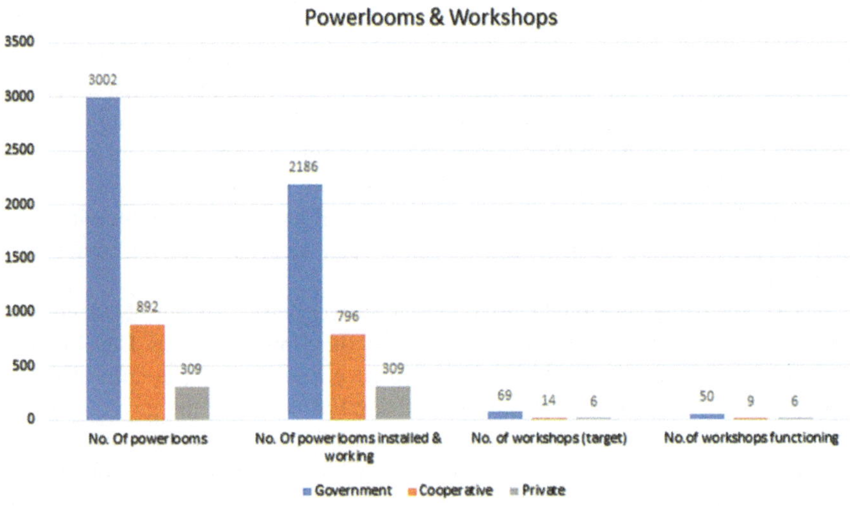

Chart 1 Particulars relating to the decentralized power loom workshop [30]

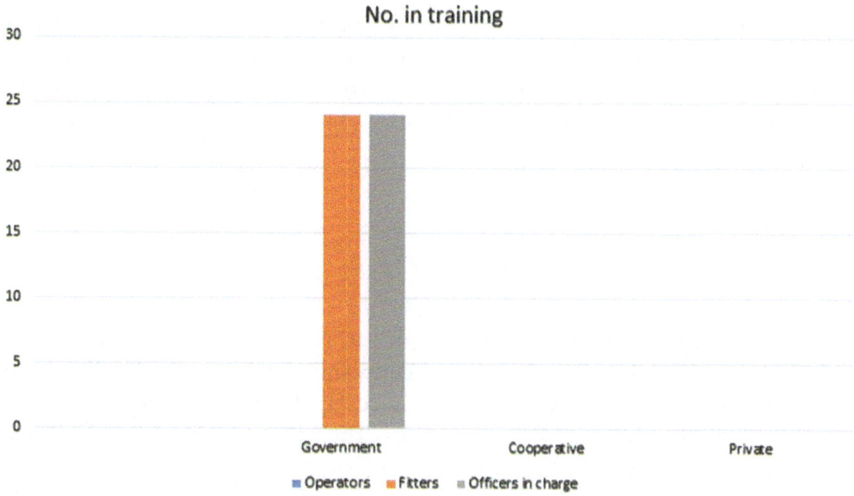

Chart 2 Training [30]

For this, young people aged 15–25 were recruited and given six months of on-the-job training. Government workshops, cooperative service workshops, and private workshops are three [30] (Chart 3).

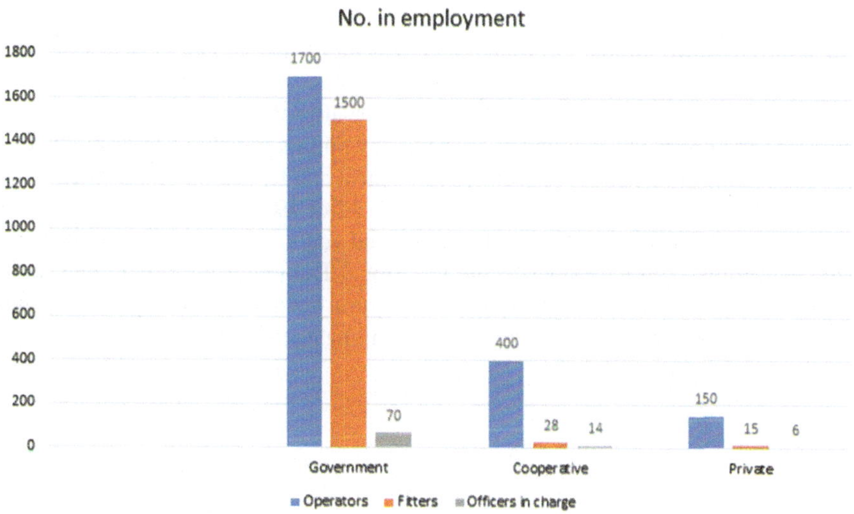

Chart 3 Employment opportunities [30]

The economic implications of these programs have not received enough attention. Even if the person works full time, most have grown merely to provide a small supplementary income to the rural family. There was no successful cooperation between the handloom textile industry and the decentralized power loom and textile factories.

5.1 National Apprentice Board 1971

The National Apprenticeship Act No 49 of 1971 provided for establishing the National Apprenticeship Board and Apprenticeship fund to provide all matters connected with its functioning. The apprenticeship fund was launched with the support of both the public and private sectors. The apprenticeship fund also funds the Katubedda Technical College in Moratuwa (now known as the University of Moratuwa) and the Hardy Institute in Ampara, offering a one-year practical training program. There were 400–500 students who obtained weaving and spinning instruction from 16 different trades. The board's objective is to formulate, implement, and supervise a scheme of training to cover each category of apprentices concerning each training and determine the period of training level of training to be undergone by each category. To determine the trade tests/issue certificates to determine the hours and conditions to work, leave holidays [30].

5.2 Political Intervention on Establishment of Government Institutions to Promote Vocational Training

The return to power of the "Sri Lanka Freedom Party" (SLFP) in 1994, whose core electorate was the rural population of the Singhalese part of the country, resulted in a dramatic policy change [42]. Only one year after the elections, the new labor minister accomplished a significant milestone by putting together a large number of training organizations under his Ministry's control and by setting up the Vocational Training Authority (VTA) in 1995 under act no 12 of 1995, which began providing vocational training programs to rural unemployed youth in general, along with some basic education. Within three years, the Vocational Training Authority opened about 200 rural vocational training courses, Centers (rural vocational training centers) and thus became Sri Lanka's largest enrollment training agency [27].

6 Introduction to the Textile Industry and Human Capital Supply

Human capital has been known as a significant factor in individual productivity for a long time [43]. More recently, it has been increasingly recognized as a factor shaping industries' competitiveness [2, 17, 18, 26, 36, 49]. Also, it is a significant element that affects the recognition of opportunity in any industry. Human capital is the expertise of an entrepreneur, which he may have obtained through education, training, or experience. In particular, Kirzner's principles of entrepreneurial alertness argue that the process of entrepreneurial exploration is correlated with the interpretation framework of the actor or the stock of knowledge derived from daily life experiences [52]. Yu claims that "everyone is overwhelmed by chances for traditional economics, but they only remain once they have been observed" is a paradox. However, by taking a subjective point of view, he states that an entrepreneur's conceptual framework plays a pivotal role in recognizing benefits and not knowledge itself. However, the way people apply knowledge is essential to recognizing opportunities.

7 Government and Non-Government Textile Education Offering Universities and Institutes

Three large companies in Sri Lanka have been recognized as collaborating to manage human capital for the newly reorganized clothing and textile industry. Those are the,

1. Apparel and textile industry, represented by the Forum of the Joint Apparel Association (JAAF),
2. The institutions of education and training (Universities and other institutions)
3. Government, trade, and industry

The Education and Textile Consortium established the textile sector expansion plan. The World Trade Organization confirmed in 2002 that it would remove the quota limitations on clothing and textiles by 2005 and remove China's quota constraints by 2008. JAAF was organized by a group of clothing and textile companies in Sri Lanka to address this major change in the global export of clothing and textile products [11]. The COAT structure formed an equal relationship for each of the three areas of shared concern for the human resource development for the clothing and textile industry, such as the JAAF, the University, and the government [11]. A bridging dysfunctional transition group was initiated from the Government stakeholder group. The Sri Lankan Government, through Parliament ACT, No. 12 of 2009, jointly recognized its clothing and textile industrial vocational training resources under the Ministry of Industry and Commerce. Its specular highlights for education, such as people and technological innovation, are key components of any successful venture, whether in the private sector or the government sector [11]. Textile is shown to facilitate the design and lead it by providing enhanced technological advances that can influence and inspire the thoughts and ideas of the learners. Fashion learners have become more knowledgeable about fitness for purpose and have sophisticated their comprehension of the use of materials. Fashion is no longer just about design style lines and colors, for the new generation of fashion designers and clothing designers, so much about functionality and, therefore, textile education for fashion learners is more important than ever to ensure that our learners mature into well-positioned designers who are entirely conversant with and equipped for the cutting-edge technological advancement to them [35].

8 Evolution of the Textile Education Towards 21st Century

8.1 Beginning of the Export Oriented Garment Industry in Sri Lanka

The textile and clothing (T&C) industries form a significant part of manufacturing development, employment, and trade in many developing countries. In Sri Lanka, the textile and apparel industry are the leading manufacturing sub-sector accounting for 67% of overall exports. Commercial manufacturing of clothing started in Sri Lanka in the early 1960s. The industry began with 19 firms in 1973 and, in 1986, the textile and apparel sector overcame plantation exports and remained the main contributor to export growth ([50]: 4). Sub-sequently, the importance of the industry for the export orientation of the local economy emerged with the establishment of an open economic policy in 1977. Over the years, the contribution of the industry to economic and human development has also been significant ([50]: 1). Government enterprises heavily influenced the textile industry—manufacturing continued to remain exclusively private sector activity, the import of the yarn and fabric needed for garment manufacturing remained under strict licensing, and the SOE-dominated household textile industry had constrained capabilities to fulfill the required specifications of the garment industry [3]. The fast expansion of the export-oriented garment industry in Sri Lanka began only after the economic liberalization reforms introduced in 1977–1979 and was reinforced by the second phase of policy changes in the first half of the 1990s. Since 1986, it has become the most significant export industry in Sri Lanka. Initial policy changes included the setting up of an effective Free Trade Zone (FTZ) strategy. Reform of trade policies was carried out in the early 1990s (including abolishing import duties on textiles and yarn). At the same time, it has been the country's largest net foreign exchange earner since 1992. The global news media titled Sri Lanka is a new investment center in Asia in the immediate post-reform years. Many of the growing regional blocks such as India, Pakistan, Bangladesh, and Cambodia are undermining Sri Lanka's market share in the world, impacting Sri Lanka's textile and clothing industry directly. Trade treaties have played a pivotal role in establishing the clothing industry in the coming years, and this situation could be further intensified. Sri Lanka signed an agreement with the EU in 2000. In January 2001, the EU excluded quota constraints on exporting textiles and apparel to EU countries. The total value of the sector's exports amounted to USD 2.424 million, representing 52% of the total export earnings in 2002. In 2002, the Gross Domestic Product (GDP) commitment increased to 5.3%. Sri Lanka is the EU's 20th biggest supplier of textiles and clothing, while the EU is the second-largest producer of Sri Lankan textiles and apparel. In mutual recognition, Sri Lanka has consented to enforce 0% import duties on raw materials, 5% on yarns and textiles, 10% on textiles, and 17.5% on garments (current tariffs are 0, 0, 25, and 10%, respectively) [9].

8.2 *Apparel Industry and Textile Sector: Opportunities in the Industry*

The textile and clothing industry is among the ancient, biggest, and most worldwide industries globally. On imported inputs, the Sri Lankan apparel industry has been established. The government's 1983–1987 Public Investment Program document has been reported as troubling the industry's strong reliance on imported inputs that account for nearly 70% of the total export value. Local exporters have no control over the components of such large imports for their exports. Local exporters have no control over the components of such large imports for their exports. Local producers to this industry have not grown considerably, despite the rapid growth of the clothing industry. Indeed, in the last 30 years, the clothing industry has had the least value-added ratio of output to between 63–70% of the value of total clothing production, a proportion so high that some talking heads have referred to the clothing industry as a glamorized tailor. Since 1988, the fabric input for the apparel industry has changed drastically. The domestic fabric supply cover ratio fell from 36% in 1989 to almost 15% in 1997.

The local textile industry can supply knitted fabrics at reasonable prices and quality, but the apparel industries are almost dependent on foreign deliveries for woven fabrics [33]. It is a conventional start-up business sector for countries engaged in export-oriented industrialization and is labor-intensive. Textiles and clothing offer an array of experiences, including employment opportunities for untrained workers in developing countries [24]. Sri Lanka's total exports for 0. 69% of total world trade [50]. Average textile imports expenses have accounted for more than 50% of T&A export earnings and a fifth of total imports since 1990 [50].

9 Issues Relate to Existing Textile Education

In the new millennium, the sustainable application of textile has become an issue in the industry. Stakeholders are encouraged to invest in their sustainable applications, which require education research and application [15, 19, 22]. In the context of increasing global textile demand, it shows that synthetic and more 'natural' textiles present substantial challenges to the environmental quality and ecological health of freshwaters worldwide [19, 41].

The global textile and clothing industry is primarily regulated by the Multi-Fiber Agreement (MFA) of 1974. The Agreement sets out guidelines for the Member States on T&A trading and making deals bilateral treaties and a framework for quantitative restrictions on trade management ([50]: 3). To maintain a successful Future of International Development Cooperation is suggested a national development vision [15]. Political Instability and trade and fiscal deficit have been identified as development progress and challenges in the new millennium. Industry leaders have identified that specializing in producing specific product categories will enable Sri Lankan manufacturers to streamline their production processes and develop skills that will ultimately make them worldwide experts in their chosen product categories [25]. Sri Lanka's workforce has achieved demand throughout its career. Therefore, it recommends that the highly skilled and educated workforce in Sri Lanka also lend itself to quicker up-skilling and workplace mobility than other countries, and skilled workers are in demand from other countries' apparel industries though it now faces competition from Asia [15]. Lessons may be learned from past experiences and need to enforce the education to find better pathways.

10 Summary

Learning is at the heart of human life, and it is the instrument through which cultures are endured and evolve. This chapter discusses the education system that developed with the traditional practices of knowledge transmission, teaching, and learning mechanisms, and the role of social learning in textile craft education in determining what is transmitted. Textile learning as a craft, a form of self-expression, and a method of learning a specific skill display a unique cultural phenomenon that reflects tradition and culture. Understanding cultural differences is easier when textiles are viewed as a craft. However, as generic human skills become more important in learning, it becomes increasingly important to transcend traditional subject divisions by grouping learning objectives under common themes. Learning is important in distinguishing between inferior and superior professions. Excellent jobs meant to require higher intellectual capacities may be assigned to those who had more learning opportunities.

On the other hand, learning can be restricted to 'learning curves' for those in lower-paying jobs. Workers observe and achieve speed and quality by repetitive task processing, which results in manual labor in production systems. Learning embedded in the traditional society of Sri Lanka is treated as a by-product of doing (learning–by–doing).

Textile education has driven through two pathways in Sri Lanka. The first one is the traditional educational system. The system was merely rooted in civil society; occupation and life became one. This is a domestic industry. From growing cotton to production and the selling has been carried out by the craftsman. The employee and the employer were the craftsmen themselves, and the learning has been handed down from generation to generation. The learning pattern remains and is sustainable in Sri Lanka even today. The end of the Cultural Revolution and the introduction of country-wide economic reforms, which saw the return of private enterprise, marked the second half of the 1970s as a period of change.

In today's modern market, consumer demands and needs are constantly changing. New processes and steps are needed to keep up with these changes. Modern textile education centers with state-of-the-art facilities will be evolved in Sri Lanka to cater to these demands.

References

1. Act no 20 of 1992 Parliament of the democratic socialist republic of Sri Lanka, tertiary and vocational education. Published as a supplement to part II of the Gazette of the Democratic Socialist Republic of Sri Lanka
2. AlQershi N (2020) Strategic thinking, strategic planning, strategic innovation and the performance of SMEs: the mediating role of human capital. Manage Sci Lett 11(3):1003–1012. https://doi.org/10.5267/j.msl.2020.9.042
3. Athukorala PC, Ekanayake R (2018) Repositioning in the global apparel value chain in the post-MFA era: strategic issues and evidence from Sri Lanka. Dev Policy Rev 36:O247–O269
4. Bandaranayake S, Dewaraja L, Silva R, Wimalaratne KDG (1990) Sri Lanka and the silk road of the sea. The Sri Lanka National Commission for Unesco and the Central Cultural Fund
5. Coomaraswamy A (1913) Art and crafts of India and Ceylon. Today and Tomorrow's Printers and Publishers, New Delhi
6. Coomaraswamy A (1956) Meadival Sinhalese art. Published by the National Museum, Sri Lanka
7. Coomaraswamy R (2012) The 1972 Republican Constitution in the Postcolonial Constitutional Evolution of Sri Lanka'. The Sri Lankan Republic act 40:126–144
8. Coomaraswamy AK (1907) Notes on Painting, Dyeing, Lacework, Dumbara mats and Paper in Ceylon, J Royal Asiatic Soc (C. B), Vol. XIX, No. 58
9. Dheerasinghe R (2003) Garment industry in Sri Lanka challenges, prospects and strategies. Staff Stud 33(1–2):31–72
10. Ellawala H (1969) Social History of Early Ceylon. Published by the Department of Cultural Affairs, Colombo
11. Erakovich R, Sipovac D, Hart S, Bruce RR (2013) Sri Lanka textile industry's public private partnership project: a case study. Int J Public Sect Manage
12. Geiger W (1960) Culture of Ceylon in medieval times. In: Bechert H (ed) Otto Harrassowitz, Wiesbaden
13. Geiger W (1953) Culavamsa-part II. Colombo: The Ceylon Government Information Department
14. Geiger W (1953) Mahavamsa. Colombo: The Ceylon Government Information Department.
15. Gu J, Chua PF, Trebs L (2020) The future of international development cooperation: case studies of Sri Lanka and Kenya.
16. Gunatillake G (1999) Development policy regimes. In: Lakshman WD, Tisdell CA (eds) Facets of development of Sri Lanka since independence: socio-economic, political, economic, scientific and cultural, Brisbane, pp 177–209
17. Hamadamin HH, Atan T (2019) The impact of strategic human resource management practices on competitive advantage sustainability: the mediation of human capital development and employee commitment. Sustainability 11(20):5782

18. Hasanbeigi A, Price L (2015) A technical review of emerging technologies for energy and water efficiency and pollution reduction in the textile industry. J Clean Prod 95:30–44
19. Islam MM, Perry P, Gill S (2020) Mapping environmentally sustainable practices in textiles, apparel and fashion industries: a systematic literature review. J Fashion Mark Manage Int J. https://doi.org/10.1108/JFMM-07-2020-0130
20. Jayawardana C (2020) History of machines and manufacturing in Sri Lanka. J Royal Asiatic Soc of Sri Lanka, Science and Technology in Ancient Sri Lanka (STASL) Publication Series - Volume III. ISBN 978-624-5697-02-1
21. Jayasuriya JE (1969) Education in Ceylon before and after independence 1939–1969. Colombo Associated Educational Publications, Sri Lanka
22. Jia F, Yin S, Chen L, Chen X (2020) Circular economy in textile and apparel industry: a systematic literature review. J Clean Prod 120728
23. Kannangara CW (1944) Report on education of Ceylon: report on special committee meeting, XXIV. Deepani Publishers, Sri Lanka, Nugegoda
24. Keane J, Te Velde DW (2008) The role of textile and clothing industries in growth and development strategies. Overseas Dev Inst 7
25. Kelegama S (2009) Ready-made garment exports from Sri Lanka. J Contemp Asia 39(4):579–596. https://doi.org/10.1080/00472330903076875
26. Kim KY, Pathak S, Werner S (2015) When do international human capital enhancing practices benefit the bottom line? An ability, motivation, and opportunity perspective. J Int Bus Stud 46(7):784–805. https://doi.org/10.1057/jibs.2015.10
27. Leeratne PM (2001) Study on the relationship between education, training and employment. ILO Area Office, Colombo
28. Mahalingam S (2000) A brief history of the faculty of engineering. Memories of an engineering faculty (ed) Faculty of Engineering, University of Peradeniya, Peradeniya, pp 1–16
29. Maurer M (2012) Structural elaboration of technical and vocational education and training systems in developing countries: the cases of Sri Lanka and Bangladesh. Comp Educ 48(4):487–503. https://doi.org/10.1080/03050068.2012.702011
30. Marga Institute (1974) Non-formal education in Sri Lanka for ICED/UNICEF. Marga Institute, PO Box 601, Colombo
31. Ministry of Textile Industries (1987) Review of activities 1981–1986. Ministry of Textile Industries, Colombo
32. Paranavitana S (ed) (1970) Inscriptions of Ceylon, vol I. Department of Archeology, Colombo
33. Pararajasingham E (2006) Textile history of Sri Lanka. Paranan Associate Private Limited, 403 1/1, Galle Road, Wellawatta. Colombo
34. Perera BJ (1952) The foreign trade and commerce of ancient Ceylon, II—ancient Ceylon and its trade with India. Ceylon History J I(3).
35. Power J (2007) An examination into textile education by active research. The Textile Institute
36. Prasetyo PE, Kistanti NR (2020) Human capital, institutional economics and entrepreneurship as a driver for quality and sustainable economic growth. Entrepreneurship Sustain Issues 7(4):2575. https://doi.org/10.9770/jesi.2020.7.4(1)
37. Sharma KL (1988) Society and polity in modern Sri Lanka. South Asian Publishers Pvt. Ltd., New Delhi
38. Shasthri NKA (1955) The history of South India: from prehistoric time to the fall of Vijayanagar. Oxford University Press, Geoffrey Cumberlege
39. De Silva CR, De Silva D (1986) Education in Sri Lanka 1948–1985: an analysis of the structure and a critical survey of literature. The Asian Foundation, Colombo
40. De Silva KM (2005) (The first Sri Lankan Edition) A History of Sri Lanka.Colombo: Vijitha Yapa Publications. Colombo
41. Stone C, Windsor FM, Munday M, Durance I (2020) Natural or synthetic—how global trends in textile usage threaten freshwater environments. Sci Total Environ 718:134689. https://doi.org/10.1016/j.scitotenv.2019.134689
42. Sumathipala KHM (1968) History of education in Ceylon 1796–1965. Thisara Publishers, Dehiwala, Sri Lanka

43. Teixeira A (2002) On the link between human capital and firm performance. A theoretical and empirical survey
44. Tertiary and Vocational Education Commission (1992) The first development plan for tertiary and vocational education in Sri Lanka. Colombo: tertiary and vocational education commission
45. University of Ceylon, History of Ceylon (1960) Colombo, vol II
46. Vangeyzel GE (2008) Traditional textiles in the Colombo national museum. Published by National Museum Colombo
47. Veragoda A (ed) (2005) Pūjāvaliya. Ratna Publishers, Colombo, Sri Lanka
48. Vitharana V (1999) Sri Lanka; the geographical vision. Department of Educational Publications, Sri Lanka
49. Vogel J (2015) The two faces of R&D and human capital: evidence from Western European regions. Pap Reg Sci 94(3):525–551. https://doi.org/10.1111/pirs.12084
50. Weeraratne B (2004) Textile and apparel industry in Sri Lanka: an empirical analysis in a globalization setting. East–West Center
51. Wickramasinghe N (2003) Dressing the colonised body. Orient Longmen Private Limited, India
52. Yu TF (2001) Entrepreneurial alertness and discovery. Rev Austrian Econ 14(1):47–63

U. G. Samudrika Wijayapala holds a Bacholor's Degree in Chemical & Process Engineering , at the University of Moratuwa, Sri Lanka. She is a Professor of Textile and Apparel Engineering at the Department of Textile & Apparel Engineering, University of Moratuwa is a qualified Charted Engineer. She is an author of 2 scholarly books, Exploration of Newer Natural Dye Yielding Plants to Sri Lankan Textile Industry and History of Natural Dyes in Sri Lanka. She is an active researcher in the field of natural dyes. She is a co- author of the text book "The Buddhist Robe": The path pointing to natural dyes and possible establishment of a natural dyeing industry in the apparel sector of Sri Lanka with special reference on the Buddhist Robe, published in 2019.

Ajith de Alwis is a Professor in Chemical and Process Engineering at the Department of Chemical and Process Engineering, University of Moratuwa and is currently the Dean, Faculty of Graduate Studies at the same University. He obtained Bachelor's degree from the University of Moratuwa and Master's degree from Postgraduate Institute of Management, University of Sri Jayawardenepura and PhD from the University of Cambridge, UK. He is a member of American Institute of Chemical Engineers USA and an Associate Member of Institution of Chemical Engineeers UK.

Gayathri Madubhani Ranathunga is a Professor of Textile and Apparel Engineering in the Fashion Design and Product Development Degree Course, Department of Textile and Apparel Engineering, Faculty of Engineering, University of Moratuwa, Sri Lanka. She holds a Bachelor Honour's Degree in Fine Arts from the Faculty of Arts & Humanities, University of Kelaniya, Sri Lanka. Prof. Gayathri is authored 2 scholarly books titled Fashions of Royalty: The Kandyan Kingdom of Sri Lanka and Art and Tradition of Sri Lanka Volume 01 Music of Sri Lanka. She is a co- author of the text book "The Buddhist Robe": The path pointing to natural dyes and possible establishment of a natural dyeing industry in the apparel sector of Sri Lanka with special reference on the Buddhist Robe, published in 2019.

Virajini Medagedara Karunaratne is a Professor of Textile and Apparel Engineering in the in the Fashion Design and Product Development Degree Course, Department of Textile and Apparel Engineering, Faculty of Engineering, University of Moratuwa, Sri Lanka. She holds a Bachelor Honour's Degree in Fine Arts from the Faculty of Arts, University of Peradeniya, Sri Lanka. Prof. Virajini is authored 2 scholarly books titled Fashion of Royalty: Kotte Kingdom of Sri Lanka and Art and Tradition of Sri Lanka Volume 02: Dance of Sri Lanka. She is an active researcher in the field of Fashion and Textile Design. She is a co- author of the text book "The Buddhist Robe": The path pointing to natural dyes and possible establishment of a natural dyeing industry in the apparel sector of Sri Lanka with special reference on the Buddhist Robe, published in 2019.

Textiles as Heritage in the Maldives

Aminath Abdulla and Marcella Schmidt di Friedberg

Abstract Although primary school enrollment rates are extraordinarily high in the Maldives, higher education institutions are relatively recent and have only seen rapid expansion over the past few years. In the field of textiles, sewing and tailoring remained extremely local, and, at present, there are still no university courses. However, the situation is evolving rapidly, with many new projects and recent initiatives by government, state-owned and private enterprises, private individuals, and NGOs. Formal textile education remains marginal, especially for traditional products. For this reason, the chapter focuses on traditional textile arts as heritage assets at risk of disappearance. Traditional crafts are perhaps the most tangible manifestation of intangible cultural heritage. Over the centuries, the Maldivian people have developed a body of traditional knowledge suited to a fragile environment dominated by the sea and a scarcity of land resources. Ancient traditions involving the arts of weaving and embroidery have always been of particular importance in the country. Maldivians have used local resources (reeds, coconut leaves, screwpine leaves) to weave decorative mats and other items, such as thatch, rope, and sails. This chapter focuses on two products—mats (*thundu kunaa*) and lace (*kasabu libaas*)—which we have chosen to illustrate the main characteristics of Maldivian textile art. These distinctive traditional skills are now at risk of disappearing with older generations and, in some cases, now owe their survival to the recovery of craftsmanship and memory for the tourist industry. Traditional weaving has also been revisited and adapted to contemporary fashion.

Keywords Textile education · Heritage · Tourism · Traditional knowledge · Decorative mats · Lace

---

A. Abdulla
Maldives Authentic Crafts Cooperative Society (MACCS), Malé, Maldives

M. Schmidt di Friedberg (✉)
Department of Human Sciences for Education "Riccardo Massa", University of Milano-Bicocca, Milan, Italy
e-mail: marcella.schmidt@unimib.it

X. Yan et al. (eds.), *Textile and Fashion Education Internationalization*, Textile Science and Clothing Technology, https://doi.org/10.1007/978-981-16-8854-6_8

1 Introduction

The Republic of Maldives is an island state (298 km^2) in the Indian Ocean, to the southwest of the Deccan Peninsula. It stretches for 750 km from the island of Lakshadweep in the north to that of Chagos in the south. Of its 1192 coral islands, which form 20 natural atolls, 188 are inhabited (2017) [1], and 154, plus 12 under construction [2], are tourist resorts, while others are used for agricultural and productive purposes. One-third of the resident population (total 378,114)[1] lives in Malé, which is about 5.8 km^2. Only nine islands are naturally larger than two square kilometers, and on many, land reclamation projects are underway to increase surface area. In recent decades, the Republic has undergone rapid and complex changes to its political, socio-economic, and cultural structure, driven by a steady increase in foreign investment in the tourism sector and new consumption models. Tourism has become the country's main productive activity, contributing 21% of GDP in 2019; tourist admissions increased from 1,234,248 in 2015 to 1,702,887 in 2019, implying an annual growth rate of 15% [3, p. 4]. Today, the country internationally embodies the Western ideal of a luxury tourist paradise. This model, which is expressed in the slogans *"the sunny side of life", "no news no shoes", "Maldives, your Perfect Escape"*, reflects the "desire to locate the imagined earthly paradise in the real geographical present" [4, p. 44]. This idyllic image has remained unchanged across the four Tourism Master Plans (T.M.P.s) that have regulated tourism development in the nation since 1982. As observed in the fourth T.M.P. (2013–2017), from the outset: "Coral islands were already the prime attraction. 'A relaxing holiday on the beach and in the water' was the main product. The Maldives was to be positioned as the place to realize the dream of a holiday on an exotic uninhabited tropical island" [5, p. 120]. Despite an initial clear separation between tourists and inhabitants, based on the "one island one resort" principle [6], the T.M.P.s reflect a progressive drive to involve local communities in the tourism sector and leverage the value of the country's cultural heritage and national handicrafts. In 2TMP (1996–2005), Maldivian heritage and culture were presented as an added value for tourism, preserved and showcased; in 2010, the new National Museum of Male was inaugurated.

Nevertheless, as a means of promoting culture, protecting heritage, and bringing economic opportunities to the inhabited islands, local tourism has been developing very slowly, and the government is well aware of this. As pointed out in the fourth TMP: "Maldives tourists come and will continue to come primarily for beach and marine activities. Taking visitors to see historical monuments will never be a prime motivator of a visit to the Maldives" [5, p. 124]. Recently, in his presentation of the upcoming 5TMP (2020–2025), President Solih emphasized the importance of linking the country's culture and heritage more closely with the tourism industry to strengthen an already established tourism identity. The new plan is inspired by the principles and practices of sustainable tourism. It reflects a commitment to seek out new ways to conserve Maldivian cultural and natural heritage: "The Maldives Fifth Sustainable Tourism Master Plan will define strategies and activities and provide

[1] Population registered by 31 December 2017, Statistical Yearbook of the Maldives, 2018.

directions for further development of Maldives' tourism sector to promote development based on sustainable tourism principles and practices. The objectives are developing environmentally sensitive business operations, supporting the protection of cultural and natural heritage, and tangible economic and social benefits to locals" [7, p. 31] (Fig. 1).

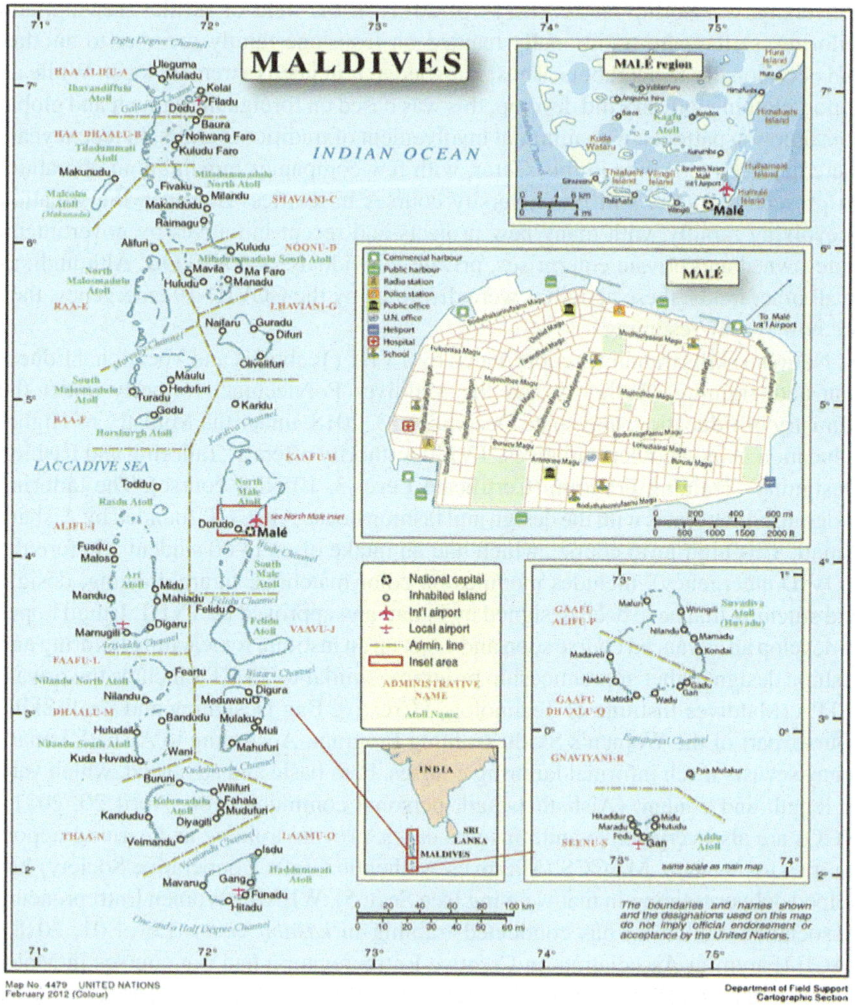

Fig. 1 Maldives map. *Source* U.N., Dept. of Field Support, Geography Section, 2012

2 Textile's Education in the Maldives

Although primary school enrollment rates are extraordinarily high in the Maldives, higher education institutions are relatively recent and have only seen rapid expansion over the past few years. The National University Act was approved in January 2011. Subsequently, the country's first third-level institution, the Maldives National University, was set up on February 15, 2011. In the field of textiles, sewing, and tailoring, skills were traditionally handed on from one family member to another and occasionally to neighbors. Thus, knowledge remained extremely local. While an export clothing industry did develop, this was based on foreign investment and global consumption patterns, with minimal involvement of traditional crafts. In recent years there has been a decline in this sector, with few companies remaining in operation. At present, there are still no university courses in textiles. However, the situation is evolving rapidly, with many new projects and recent initiatives by government, state-owned and private enterprises, private individuals, and N.G.O.s. Although, as in all other fields, these activities were disrupted by the COVID-19 emergency, they are now slowly resuming.

National skills training is coordinated by TVET (Technical and Vocational Education and Training) Authority and the Maldives Polytechnic, formerly under the Ministry of Education, and since December 13, 2018, under the Ministry of Higher Education. Among other initiatives, TVET Authority offers a "Tailoring and Fashion Designing" Training Program (Certificate Level 3, 105-day course). The tailoring program collaborates with the design and fashion house AICCET, founded by Aishath Ismail. This high-level course, which had an intake of up to 60 students (before the COVID emergency), includes modules on color matching, figure drawing, design, and stitching that have been designed by Ismail and approved by TVET. Ismail hopes to develop an advanced course soon and to set up an institute for teaching tailoring and fashion design. Other apprenticeship programs similar to TVET's include the private MIT's (Maldives Institute of Technology) Creative Bag Designing and Craft Skills course, part of the Women's Skills Training Program. According to Aishath Ismail, many sewists teach informal tailoring courses, both basic and advanced, which vary in length and content (Aishath Ismail, personal communication, April 20, 2021). NGOs are also very active and, in some cases, provide training and earning opportunities for women. MACCS (Maldives Authentic Crafts Cooperative Society) has helped deliver training in mat weaving (see Sect. 5). WEAM (Women Entrepreneurs Association Maldives) has conducted training in *kasabu viyun* (Level 01, 2018). FACE (Faiymini Association on Creative Entrepreneurs) had run courses in Male' and Laamu atoll (35 students trained before the COVID emergency) and received an award at FANNU EXPO 2019.

An interesting development in the private sector is the *Kasabu by Hidha* brand created by award-winning artist Hidhana Latheef. The brand's mission is to revitalize the tradition of *kasabu* weaving, which was traditionally used in making the Maldivian garment for women, *dhivehi libaas* (see Sect. 6), and today is employed in the production of modern fashion accessories or decorative art. The items produced are mainly sold through social networks. After completing a basic weaving course, beginning in 2017, and a period of apprenticeship, Hidhana gave up her previous job to devote herself full-time to preserving this traditional craft and cultural heritage asset and keeping it alive for future generations by educating others. She now holds courses in her home for a maximum of five students at a time and provides all the necessary materials. Classes cover both weaving techniques and the history of the weaving tradition. Introductory courses cover eight types of braid (*hiru* and *bagiyaa*), while advanced courses present more complex designs (*rodhigandu, thundu, and badhala hiru*) (see Sect. 6). At the moment, the greatest challenges concern classroom space and sourcing raw materials (Metallic yarn/*kasabu*, Metallic ribbon/*baadhala*). Students come from various backgrounds (over the past year, they have included architects, researchers, designers, anthropology students, small business owners, etc.). Interest in these classes has mainly been displayed in Male' and a few other islands. *Kasabu by Hidha* also offers a free introductory course for children with special needs, at least three students per year (Fig. 2).

Hidhana reports positive feedback from the participants in her training courses and a growing interest in learning traditional weaving crafts. *Kasabu by Hidha* aspires to introduce this intricate art to a global audience and wishes *kasabu boavalhu* to be recognized as part of the unique cultural identity of the Maldives: "The future for this craft is promising, as I can clearly see a small but committed group of girls building a great community of weavers, artists, designers, and teachers of this tradition" (Hidhana Latheef, personal communication, April 28, 2021).

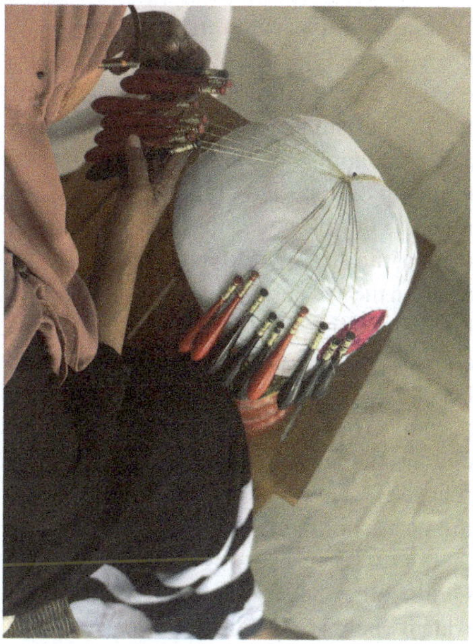

Fig. 2 Hidhana Latheef at work. Photo: Fathmath Sana, March 2018

Textiles education is still primarily private and lacks formal recognition at the national level. As a result, some individuals go abroad (Sri Lanka, Thailand) to complete diploma or degree courses in international fashion design. Along the lines of *Kasabu by Hidha*, we may note, among others, the successful career of fashion designer Noora Rasheed, who won the prize for costume design at the 2019 Maldivian Filmfare Awards. Abudy Maldives, an award-winning local artist, is famous for her intricate thread painting and decorative skills. She has also taught craftwork courses sponsored by the Ministry of Arts, Culture, and Heritage. Jeeza Mohamed, after completing a fashion design and dressmaking course in Thailand at Kalwin International, worked in these fields for over 25 years and currently teaches dressmaking courses. She has a small business called Wincey Design and has been involved in various projects, such as producing uniforms for MNDF (Maldives National Defense Force). She believes there is little demand for fashion design courses; by comparison, those who train in tailoring and dressmaking can easily find employment sewing clothes upon completing their course (Jeeza Mohamed, personal communication, May 2, 2021). Formal textile education remains marginal in the Maldives, especially for traditional products, despite the projects just outlined. Thus, pending new developments, we decided to focus this chapter on traditional textile arts as heritage assets at risk of disappearance.

3 Tourism, Heritage, and Crafts

Within the rapid cultural, social, political, and economic transformation affecting the Maldives, the protection of cultural heritage and natural resources has recently become a national priority, including tourism. While the continuous expansion of luxury tourism leaves the Western notion of the island paradise intact and represents a key economic resource, alternative representations are being sought concerning heritage: culture and crafts bear the potential to strengthen the country's national identity. Furthermore, the new and growing non-European markets (China, India, Russia, the Arab World) are inspired by a model of tourism that does not necessarily draw on the Western imaginary: "The Maldives is now a well-known brand to tourists, but the country needs to revive its arts and crafts as there are new kinds of tourists arriving from Russia, China and the Middle East who are less attracted by a pristine beach holiday and more interested in local culture and purchasing products" ([8], p. 996). The new focus on heritage has also been triggered by the emergence of climate change and rising sea levels: "Climate change is a cross-cutting development issue as it affects every aspect of the Maldivian way and livelihoods" [9, p. 10]. On the one hand, the fact that the Maldives may be at risk of "disappearance" and are therefore "to be seen now before it is too late", increases the nation's tourist appeal. On the other hand, plans for new infrastructures, land reclamation, ports, artificial islands, new settlements, and barriers against the rise of the sea and the possible death of corals due to global warming may lead to totally new scenarios in the future of Maldivian tourism. Traditional Maldivian knowledge, particularly the body of narratives and myths connected to nature and the ocean, can help to tackle the present environmental crisis. Tapping into traditional knowledge has also been helpful to conservation management and the establishment of protected marine and terrestrial areas and encouraging a new focus on the traditional use of resources. These recent change processes offer an opportunity to rethink heritage as a cross-cutting priority in the country's governance and its sustainable development. The new and active Ministry of Arts, Culture, and Heritage is committed to safeguarding objects and sites of historical importance for future generations and ensuring the documentation and protection of cultural heritage. Among ongoing initiatives, the Maldives Heritage Survey (2018) was launched to systematically inventory and document endangered cultural heritage assets and create a database of them to integrate development, sustainability, conservation, and new technologies.

Traditional crafts are perhaps the most tangible manifestation of intangible cultural heritage. The 2003 UNESCO Convention emphasizes protecting the skills and knowledge associated with crafts rather than the craft products themselves and encouraging artisans to continue to pass on these skills within their communities (https://ich.unesco.org/en/traditional-craftsmanship-00057). Island communities in the Maldives have always relied on natural resources for their sustenance and livelihoods. Traditional knowledge in the Maldives still plays a key role in many spheres of everyday life, such as fishing, agriculture, textile and garment making, food preparation, healing, health and medicine, navigation, and the interpretation of meteorology and climate [10, 11]. It can be used again today to understand the complex social and cultural transition the country is experiencing and address the environmental crisis in light of narratives and myths concerning nature and the ocean. From an environmental point of view, there is also currently a danger that increased dependence on imported products will reduce local agrobiodiversity in the long term. Maldivian handicrafts have always enjoyed a regional reputation for excellence and variety: "From the beginning, Maldives has been a melting pot of many cultures from the Indian Ocean. Its cultural heritage consists of an intangible heritage based on the unique language, performing arts, arts and crafts, cuisine and a tangible heritage based on its maritime heritage and architectural heritage from pre-Islamic and Islamic periods" [12, p. 1]. Traditionally, crafts in the Maldives have included weaving, lacquer work, jewelry, embroidery, wood and metalwork, and coir rope production, with different islands and atolls specializing in different areas. As observed by Pyrard de Laval in the seventeenth century: "such diversity is there; what is plentiful in one island is rare in another […] The people, too, in their domiciles have followed a similar rule, for the craftsmen are collected in different isles—for instance, the weavers in one, the goldsmiths, the locksmiths, the blacksmiths, the mat-weavers, the potters, the turners, and the carpenters in others; in short, their craftsmen do not mingle together; each craft has its separate island. […] Mat-weaving is confined to the islands of Huvadú, and the manufacture of cotton-cloth is carried on chiefly in Málosmadulu, Addú, and Huvadú atolls" [13, p. 114]. As in many other countries, industrialization and the introduction of cheap imported products contribute to declining such local activities [14]. Ancient traditions involving the arts of weaving and embroidery have always been of particular importance in the Maldives: yet, these distinctive traditional skills are at risk of disappearing with the older generations and, in some cases, now owe their survival to the recovery of craftsmanship and memory for the purposes of the tourist industry.

4 Traditional Knowledge and Natural Resources: Coir Rope and Thatch Weaving

Over the centuries, the Maldivian people have developed a body of traditional knowledge suited to a fragile environment dominated by the sea and a scarcity of land resources. This collection of information, rituals, and practices, which has been passed on from generation to generation, can still today offer valuable in-depth insights and guidance at the atoll and island levels and has contributed over the centuries to conserving the archipelago's landscapes and biodiversity. More generally, in-depth knowledge of plants and the products of nature has enabled the diversified use of local resources to provide food, medicine, and products for trading. Over time, Maldivians have used reeds to weave the famous Hoovadhoo atoll mats, numerous other plants to obtain color pigments, screwpine leaves for mat weaving in the Northern Maldives, coconut palm leaves for thatch weaving, and coconut palm fronds for weaving other items, such as sails, as recorded by Pyrard de Laval [13, p. 236]. The coconut palm (*Cocos nucifera*)—*dhivehi Ruh*—was declared the national tree in 1985 and is represented on the national emblem, taking first place among the plants of the Maldives. Stands of coconut, both cultivated or growing wild in the forest, have always been found in abundance on all the islands. In the past, women made coir rope (*roanu*) from coconut husks, palm leaves were tied together to make thatched roofing (*fangi*) or baskets (*Vashi*), while the mid-rib of the palm leaves was used to fashion brooms (*iloshi fathi*). Palm leaves are still used to make household items, such as hampers and mats, and decorative objects in the form of fish and birds (Fig. 3).

Throughout Maldivian history, the production of coir rope, in particular, has constituted one of the leading traditional handicrafts, driving exports from the country: "The greatest trade is in cordage" related Pyrard de Laval (idem.), while Young and Christopher added that: "The coir-yarn of this place sells higher than the ready-made rope of the continent, it being much finer, and of a higher color" [15, pp. 54–86]. Making coir rope, predominantly carried out by women, is a lengthy process: it can take two months or more to produce a single batch of ropes. Coir is separated from the coconut husk after it has been left to cure in the sea for a few weeks: "And people take from there […] *qânbar*, that is, the fibrous covering of the coconut. This is tanned in pits on the shore, beaten with mallets, and then spun by the women. Finally, ropes are made from it, which are used to bind the ships together and are exported to China, India, and Yemen; these ropes are better than those made from hemp, and with these ropes, the beams of the Indian and Yemenite ships are sewn together for the Indian Ocean has many rocks. If a ship nailed together with

Fig. 3 A traditional thatch wall is showcased on Milaidhoo Island. Photo: Mahmood Fauzy, 2018

iron nails collides with rocks, it would surely be wrecked; but a ship whose beams are sewn together with ropes is made wet and is not shattered" [16, p. 201]. Nowadays, other imported products have replaced coconut rope and fiber. Maloney documented an attempt in 1950 to set up a coir rope factory on the northern island of Malé: it soon failed due to the challenges involved in collecting coconut husk from a large number of islands [17, p. 285]. Today, the production of coconut rope continues locally as a function of tourist activities (Figs. 4 and 5).

Fig. 4 Strands of coconut husk ready for making rope in HDh. Kulhudhuffushi. Photo: Abdul Sathar 2021

Fig. 5 Coir rope made in HDh. Kulhudhuffushi. Photo: Aminath Abdulla 2021

Palms are also the raw material used for thatching, a key activity across the Maldives in earlier times, which has almost disappeared on inhabited islands with changes in construction methods and the introduction of corrugated roofs and/or limestone walls. Thatched roofs have recently been revived as a tourist attraction. Many resorts, designed to evoke a "Robinson Crusoe" model of nature, offer ultra-modern, fully-equipped huts with thatched roofs. Thatching is currently only practiced for the purposes of tourism at islands near resorts, providing a source of income for women, who have traditionally been responsible for the many manual activities involved in processing natural materials. As noted by Bell: "The domestic duties are mostly attended to by the women. They also beat out the fibers of the cocoanut husk, after it has been soaked, separate the thick from the thin, and twist them with the fingers into yarn; make mats, prepare the bread-fruit for keeping by slicing and drying it in the sun, extract oil from the nuts, spin cotton, and dyed thread for the loom" ([18], p. 61). Most notably, the women in Baa Atoll still make a livelihood from their thatching work. Traditional Maldivian thatching tools were made from natural resources available on the islands. The same tools are used today, except the wooden needle, which an aluminum needle has replaced. The following items are required for thatch weaving: coconut leaves (*Cocus Nucifera*), stiff thin coir rope, three flat needles made of Ironwood (*kuredhi*), or aluminum (some 35.5 cm long and 1.5 cm wide), and a knife or scissors (Fig. 6).

The thatching process is long and complex. Each individual section of thatch—known as a *riyanu*—is around 7 ft. long. Approximately 360 strips of coconut leaf are required to form one seven-foot section. The leaves used for thatching are primarily coconut leaves that have fallen spontaneously to the ground. Occasionally, raw coconut leaves are also used after softening them by roasting them over the fire. When cutting leaves from the stem, the weaver ensures that the edge is left whole, ensuring that the leaves are sturdy and intact. Coir rope is threaded through

Fig. 6 Image of thatch weaving using aluminum needles and coir rope. Photo: Aishath Mohamed Rasheed, 2019

the wooden needles to bind leaves together at three different leaf points, namely at the top, middle, and bottom. The length of the coir rope used will be slightly greater than the length of the section of thatch. The weaver uses three sets of wooden needles to sew the leaves tightly together, making sure they form a compact block with no spaces between them. Every tenth coconut leaf, a shorter leaf is sewn to the mid-binding point to make the thatch thicker. Quality thatching will feature about 60 such insertions per seven ft. of woven thatch. Thatching is traditionally carried out by weavers sitting on the ground, given that it requires much space. If a weaver needs to interrupt their work before it is completed, they poke their needles one-third of the way into the leaves and tie a knot at each needle (Table 1).

5 Decorative Mats: *Hau* Cultivation and the Weaving of *Thundu Kunaa* in Huvadhoo[2]

Other high-quality, traditional Maldivian handicraft products are the well-known decorative mats made of screwpine leaves and reeds. In this case, distinct specializations have developed across the different areas of the archipelago based on in-depth knowledge of local plants and nature. There is a distinction between the types of mat woven in the southern and northern areas of the Maldives. In the northern atolls, mats are mostly woven from dried screwpine leaves without any added color. In

[2] The subject of *thundu kunaa* production has already been dealt with in Schmidt di Friedberg [19], and Schmidt di Friedberg and Abdulla [20].

Table 1 Thatch weaving

Instructions for thatch weaving
1. Take two coconut leaves and leave four inches near the stem
2. Use a needle to sew the two leaves together at the four-inch mark. This is known as "Bolu Roanu". Ensure that all leaves are straight and level with each other
3. Allow 8 in. from the "Bolu Roanu" to the next binding point. This is the middle point of the leaves and is known as "Medhu Roanu"
4. Use the second needle to sew the two leaves together at this point
5. The next binding point is usually 8 in. from the "Medhu Roanu". This binding point is dependent on the length of the coconut leaf
6. Use the third needle to sew the two leaves together at this third point known as "Huguli Roanu"
7. Repeat the process with more leaves

times past, screwpine mats were woven across almost all the atolls in the North of Maldives from H.A. Atoll to Kaafu Atoll. Currently, due to tourism, the tradition is mainly carried on in Baa Atoll one of the first atolls to have many resorts, apart from the tourism development around Malé. Two kinds of screwpine leaves are used for mat weaving: both varieties—namely the *Pandanus tectorius* (*boa kashikeyo*) and *Pandanus odoratissimus* (*maa kashikeyo*)—are widely grown across the Maldives. The mats may be woven both with and without a loom. When weaving with a loom, the leaves are cut down to a centimeter in width, whereas when weaving without a loom, the leaves are used whole. In the past, these mats were used on beds; nowadays, they are often used in tourist resorts to decorate the ceilings of tourist accommodation units (Fig. 7).

Turning south, a case study of note is the production of decorative mats known as *thundu kunaa* which are exclusive to the atoll of Huvadhoo[3]: "The Maldivians display great skill and taste in the manufacture of these mats, which have acquired a reputation for harmonious design and permanency of color. They are made only in Suvadiva atoll from a grass called by the Maldivians *hau*; only three colors are used—black, dark yellow, and white, which is obtained from plants and are wonderfully lasting" [21] (Fig. 8).

The art of *thundu kunaa*, all steps practiced by women only, again draws together biodiversity and heritage by protecting a specific marshland ecosystem and producing quality handicrafts. The Huvadhoo Atoll, one of the largest natural atolls globally, is located in the southern Maldives. For administrative purposes, the atoll has been divided into two areas: Gaafu Alifu (GA) (or North Huvadhoo Atoll), whose main island is Villingili, and Gaafu Dhaalu (GDh) (or South Huvadhoo Atoll), whose main island is Thinandhoo. The marsh reed—locally known as *hau* (*Cyperus polystachyus*)—used to weave the mats grows in the southern islands' marshy areas; traditionally, weavers prefer to cultivate it on a specific island, Fiyori Island, located

[3] The current administrative name is Gaafu Atoll, the traditional name Huvadhoo (or Suvadiva) is used here.

Fig. 7 A horizontal loom (*haru*) is used to weave the *thundu kunaa* of Huvadhoo Atoll. Photo: Federica Adamoli, February 2019

Fig. 8 The reeds (*hau*) used for weaving are dried in the sun and tinged with natural black and yellow dyes or left in their original color. Photo: Aishath Niyaz, 2014

Fig. 9 Fiyori *hau* (reed) field and Fiyori reeds. Photo: Aisha Niyaz, 2014

to the southwest of the atoll. This island's special soil conditions and its wetlands, known locally as *olhu*, appear to be particularly suited to the production of hardy, durable reed plants that are light in color and not hollow inside. Fiyori's wetland system also stands out as a key ecosystem given that it hosts several species of migratory birds, most of which enjoy protected status in the Maldives. Hence, over time, the island has specialized in the production of reeds, which are cultivated and harvested with special care by the community's women. Several other islands in the atoll have attempted *hau* production in the past, but none have managed to surpass Fiyori in terms of quality or quantity [22] (Fig. 9).

With Fiyori providing the raw material, the mats are woven on the loom by the women of Gahdhoo Island, who are considered the most skilled at producing this prominent emblem of Maldivian craftsmanship, with its elaborate and ornate designs. Indeed, in Huvadhoo Atoll, a special relationship traditionally formed between the cultivation and protection of a specific plant—the Fiyori *hau*—and the ancient female art of weaving on the island of Gahdhoo, some 35 km away by sea: "These fine mats surpass anything in the same line the world over, and have justly obtained unqualified commendation" [18, p. 88]. In the course of the Maldives' interactions with the colonial powers, *kunaa* produced at Gahdhoo, up to two meters in length, were often offered as royal gifts to visiting foreign dignitaries and as tributes to the Dutch and British. A collection of beautifully decorated *thundu kunaa* are held at the Ethnography Department of the British Museum in London [23] (Fig. 10).

The reeds used for weaving are dried in the sun and then tinted with natural dyes of black or yellow or left of their original color. The dyeing process can take several days, is extremely laborious, and requires extensive knowledge of herbs. Yellow is obtained by boiling the rushes in a mixture of saltwater, *ahi* or *noni* (*Morinda citrifolia*) roots, henna, turmeric, and other plants. The particularly complex black color requires, among other steps, boiling the *hau* in a mixture of bark and of *unigas* or Beach Gardenia (*Guettarda speciosa*) leaves, *midhili* or Beach Almond (*Terminalia catappa*) leaves, palm sugar, and other ingredients, followed by a second boiling in a mixture of fermented coconut water and rusty pieces of metal. The *hau* takes on darker hues thanks to a process of oxidation. In addition to yellow and black,

Fig. 10 Parts of the mat. Photo: Aishath Aima Musthafa, 2017

white—the natural color of the dried fiber—is also used in traditional *thundu kunaa* [22, pp. 14–16]. The first step after dyeing is to prepare the horizontal loom (*haru*). Traditionally, a fiber extracted from the bark of the *dhiggaa* or sea hibiscus (*Hibiscus tiliaceus*) is used to make the warp threads: first, the fiber is left to soak in wet sand for days, then dried and made into a fine thread, whose exact characteristics depend on the size and specific features of the mat to be produced. While the designs on mats once varied according to the use for which the end product was intended and the social status of the buyer, today, the meanings and social valence of mats have primarily been lost [22] (Figs. 11 and 12).

As a Maldivian national heritage asset, the conservation of hand-woven mats is tied up with the conservation of an entire ecosystem. The skills involved in cultivating the reeds, producing the natural dyes, and weaving the mats, have been handed down from generation to generation but are now only practiced by a small number of older women and thus in danger of disappearing. Recently, cooperative systems have been springing up all over the country, intending to redevelop and market the craft and building a network of local producers. The Maldives Authentic Crafts Cooperative Society (MACCS), based in Malé, is a crafts cooperative founded in 2011 by women, including Aminath Abdulla, to support and revitalize handicrafts, in particular, the art of mat weaving in Huvadhoo. Availing of small grants programs (including UNDP's Global Environment Facility's Small Grants Programme and Mangroves for the Future), MACCS seeks to support local communities by making handicrafts a sustainable source of livelihood. The cooperative aims to restore and regenerate the native hau species in Fiyori and have the *olhu* (wetland) declared a protected area from an ecological perspective. The main challenges concern transporting reeds to

Fig. 11 Traditionally, fiber from the bark of the *dhiggaa* or sea hibiscus (*Hibiscus tiliaceus*) is used for the warp threads, which are soaked in the sand for days. Photo: Federica Adamoli, February 2019

Fig. 12 Fiber from the bark of the *dhiggaa is* dried and processed into a fine thread. Photo by: Aisha Niyaz 2016

Fig. 13 Large mat used on old coir rope beds and swings, *kalhujehi kuna* (72"X 30"). Photo: Aishath Aima Musthafa, 2020

the weaving sites: before 2017, Fiyori and Gahdhoo were not connected by public ferry, and the only solution was to bring supplies to Thinadhoo, the atoll's main island, and extend the women's entrepreneurial reach, possibly by introducing a system of intermediaries. The sale of traditional mats, now produced in new shapes and sizes for the tourist resorts and guesthouse circuits and Malé, can become a significant niche within the Maldivian economy and a new source of income for women. To stimulate demand, MACCS has developed products that are easier to use than the conventional large mats, such as traditionally woven laptop bags, coasters, and placemats. The cooperative also contributes to building up the supply chain by training more women in the requisite skills (Figs. 13 and 14).[4]

6 The Tradition of Weaving in the Maldives: *Kasabu Libaas*

Maldivian culture is relatively ancient. Archaeological and historical evidence suggests that the archipelago has been inhabited for over 2500 years [24]. Located along the sea navigation route between the Arabian Peninsula and China, the islands have always been a point of encounter for traders and travelers more generally: "The coming and going of settlers, migrants, explorers, castaways, slaves, and merchants and merchandise, religions, supernatural beings, building structures, currencies, calendars, counting and measuring systems, and scripts, as well as of old and new concerns and hazards, indicates the location of the Maldives Islands at a complex set of crossroads of major historic tidal currents of the Indian Ocean" [25, p. 15]. Knowledge about the ancient arts of weaving and making clothing on the islands

[4] In July 2019, MACCS also took some weavers to Medinipur, India, to show them a similar type of mats (called *madur* in India) and their production process. In the Medinipur district, the local weavers—men—use a similar type of reed (Cyperus Tegetum) and powder dyes. During the visit, the weavers also visited an exhibition in the village attended by all members of the community.

Fig. 14 Small mats woven for Souveniers (12“ X 11”). Photo: Aishath Aima Musthafa, 2020

and how these have developed over time has been transmitted orally and in writing in the *divehi* language over the centuries. In addition, a number of key personal accounts accessible to international audiences suggest a historically intense network of exchanges and offer, amongst other contents, information about ancient artifacts, clothing, weaving, embroidery, and jewelry on the islands. In the sixteenth chapter of his work *The Journeys* (الرحلة, *Rihla*), Ibn Battuta recounted his stay in the Maldives, where he remained for about a year and a half, beginning in 1345: “Their clothing consists of a waist-wrapper; they bind this round their middle instead of the trousers and put round their shoulders an article of clothing called *wilyân* [...]. Some put on a turban, while others wear a small kerchief” [16, p. 199]. Women, including the Queen, did not cover their heads: “Most of them wear only a waist-wrapper which covers them from their waist to the lowest part, but the remainder of their body remains uncovered. [...] The clothes of some of them consisted of a shirt (*qamis*) besides the waist-wrapper, and their shirts had short wide sleeves” (Id., p. 202). Ibn Battuta listed waist-wrappers, *wilyân*, and cotton turbans as leading export goods, alongside coconut and fish.

Two and a half centuries later, the French navigator Pyrard de Laval was shipwrecked with the sailing ship Corbin in Baa atoll in the Western Maldives. He remained in the country for five years (1602–1607), mainly on Malé, learning the language and providing, among other contributions, a detailed account of local customs that reflected, on the whole, a highly optimistic view of Maldivian society: “In short, the people are exceedingly adroit, much given to the manufacture of all kinds of things, and excelling therein, even in letters and science, according to their notions” [13, p. 106]. De Laval also offered a detailed description of traditional

Maldivian clothing: "As for their garments, see how they apparel themselves. First, the men bind all-around their private parts a broad band of cloth, lest these should be discovered in coming or going or in doing some work. Over this, they wear a small cotton cloth, dyed blue or red or other color, reaching only as far as the knees. Above this they wear a large cloth of cotton, or silk if they are at all rich and well off; this reaches to the ankles, and is and belted with a fine square kerchief, embroidered with gold and silk; this is folded in three, drawn round the loins, tied in front. Then, for braver adornment, they add a little piece of silk of divers colors, fine as crape or gauze, and short, not reaching further than the midthigh; and after all this, they gird themselves with a large silken sash, like their turbans, with pretty fringes, the ends of whom they let hang down in front" (Id., pp. 161–162). Regarding women: "first of all, they have a large colored cloth of cotton or silk, which covers them from the waist to the ankles, and serves as a petticoat. Above this, they wear a robe of taffetas, or very fine cotton, but very long, reaching to the feet, and with blue and white borders" (Id., p. 166).

In the nineteenth and twentieth centuries, other Western scholars "discovered" the Maldives, bringing back fresh information about Maldivian dress and crafts, among other topics. In 1884, survey Lieutenants Young and Christopher, of the hydrographic expedition led by Captain Robert Moresby [26], added new details on the Maldivian way of dressing: "The ordinary dress of the men consists of short drawers, with a cloth wrapped around the waist, and another about the head, the waistcloth being twisted into a knot in the front, which is supported by a string encircling the loins, The head people wear, in addition, an embroidered sash of silk or cotton about the waist, and on Fridays, when attending the grand mosque, a kind of shirt (white) reaching to the ankles, with a turban of the same color. [...] The women's habiliments consist merely of a cloth wrapped around the waist, descending to the knees, secured by a string, and a long shirt: also a cloth tied around the head" [15, pp. 58–59]. In addition to verbal descriptions of the Dhivehi attire, we also keep photos and pictures of 19th-century travelers. The German photographer Rosset visited the islands in 1885 and printed a large number of illustrations and photographs of contemporary notables and scenes from everyday life [27]. Some of the photos feature traditional clothing, including the author's collection of Maldivian garments (Fig. 15).

From the comments of the nineteenth-century authors, it is clear that textiles manufacturing was one of the main economic activities in the Maldives at that time. Bell reported that "Weaving cloth is one of the most common employments" [18, p. 61]. Cotton or *ui*, which was cultivated on only a few islands, was mainly imported: "Cotton is grown in small quantities at one of the islands, Zilla-da-Male" [15, p. 80]. According to Maloney, the weavers (men) of the southern islands "were put out of business as mill cloth became popular during the British period" [28, p. 285]. Young and Christopher added: "The principal manufactures, and the only ones in which the natives have attained any degree of skill, are those of mats and cloth". This chapter focuses precisely on these two very products—mats and cloth—which we have chosen to illustrate the main characteristics of textile art in the Maldives. We have already examined the case of mats. Young and Christopher further commented: "The former is made only at Suadiva Attol, not in great quantity. The cloth is made

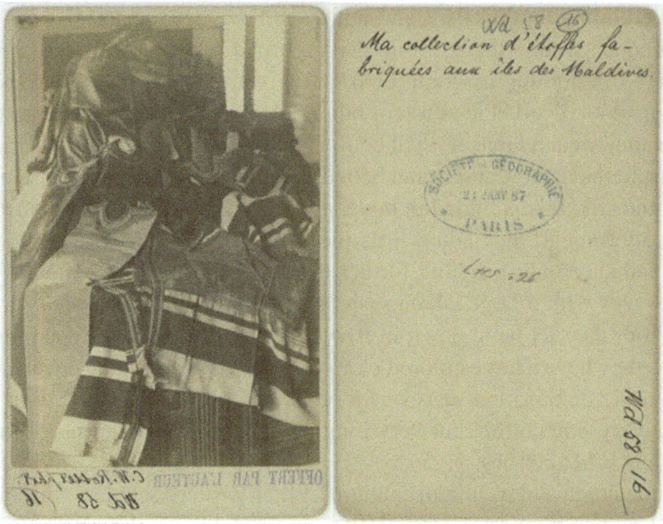

Fig. 15 Rosset, Male', Maldives 1885—my collection of textiles made in the Maldives islands. *Source* gallica.bnf.fr/Bibliothèque nationale de France

in Malusma-Attol principally, although individuals of other islands sometimes make it for their use. This article's peculiarity consists of the uniformity of the coloring of the waist and head cloths, which are very becoming being chiefly of red color, with alternate black and white stripes, each also having a neat fringe. The dyes are very excellent, particularly the red, which is extracted from a root called *Ahi* [*Morinda Citrifolia*]; the black dye is made from burnt cocoanut husks" (Id., p. 81).

In Bell's account of a few decades later, traditional Maldivian dress held in high regard thanks to a newfound appreciation of local heritage: "The ordinary dress of the men consists of short drawers (M. *haruválu*), a cloth wrapped, after the Sinhalese fashion (M. *mundu*), and a plain handkerchief twisted over the head (M. *rumá*). [...] the women wear a waistcloth, generally of native manufacture, coarse in texture, the ground of a chocolate color, relieved by black and white stripes (*M. féliya*). The upper part of the body is covered by a loose-fitting, red-colored jersey reaching the knees, short-sleeved, and edged at the neck with silver tinsel lace (M. *libás*), while round the high-worn *kondé* is twisted a handkerchief matching the *libás* in color, the tout ensemble forming a very becoming and picturesque costume. A profusion of bangles, with necklace, earrings, and other ornaments, deck the person. The use of gold trinkets is no longer rigidly restricted to women of quality" [18, p. 55]. Bell suggested that Maldivian folk attire (*feyly viyun*) came from a long tradition, with descriptions of it surviving from 1340 onwards [14, p. 10]. It was a comfortable form of dress, cut from hand-woven cotton well suited to the heat of the equatorial climate. Conventionally, the *feyli* was worn by both men and women: as a sarong with a long white sleeved shirt by men, and as an underskirt by women, under the *divehi libaas*, the traditional female costume. This type of weaving was mainly

Fig. 16 *Feyli*, made by the last Dhevadhoo weaver. Photo: Aminath Abdulla, 1998

performed by men and was practiced across the entire country. The traditional *feyli* was habitually donned during the monarchy period, up to 1932, as an official dress for special occasions (Id.). The traditional centers of *feyli* production were the islands of Dhevvahdu, Fodhdhu, and Kachchimidhu and, more recently, the island of Eydafushi [29] and that of Dhehvadhoo in Baa Atoll. Until recently, there was one last remaining person in Dhehvadhoo practicing traditional cloth weaving, but he passed away in 2010. Currently, only one person in the entire country is known for his *feyli* weaving: this person is also an older adult and does not weave for commercial purposes (Fig. 16).

The traditional Maldivian outfit for women is known as the *Dhivehi libaas*. The most characteristic part of this elegant female costume is the *boavalhu* (a set of colored lace bands positioned at the neckline) woven using fine golden and silver thread imported from India. *Kasabu*[5] *boavalhu* comprises a colorful sequence of thin lace bands that extends approximately 4.5 in. from the neckline of the dress. The same pattern is reproduced on both sides of the dress. It requires making particular use of bobbins, and women mainly carry out the weaving work. The *Dhivehi libaas* is a long-sleeved dress that falls to one inch below the knee. It is worn with a black sarong (*feyli*) underneath. The dress may be plain or embellished with *kasabu* by those who can afford it. Its use was more widespread in the first half of the twentieth century and was only worn by middle- and upper-class women prior to World War II. However, during the war, it became difficult to source materials for the *kasabu boavalhu*. Thus, printed cotton fabric was adopted in place of the original woven *kasabu lace bands*. This new design became known as the *fothi boavalhu*. Women from the southern Maldives—mainly Huvadhoo and Addu—wear *kasabu boavalhu*

[5] This term probably originated in Bengal, given that a similar type of dress named *kasavu* is to be found there [30].

Fig. 17 A close up of woven laces fixed together on a *libaas*. Photo: Aishath Aima Musthafa 2017

more frequently than do their counterparts in other areas of the country. Nowadays, women wear dresses with decorative lace bands known as *kasabu libaas* to official functions. The *boavalhu* is designed to run all around the *libaas*. There are three main types of woven lace band, namely *bagiya*, *hiru*, and *rodhigandu* (Fig. 17):

Bagiya introduces two alternative styles of the weave. The *vah bagiya* refers to the last two—distinctively woven—lace bands on a *kasababu boavalhu*. The last band is sometimes replaced with a *thundu* (see image) to make the *boavalhu* more ornate. A *thundu* is a band that is woven in a sawtooth pattern. The other distinctively *bagiya* style is applied to the band immediately beneath the *mathee kairi* (the band at the neckline itself).

Hiru comprises four different weaves of variable depth, which are labeled *math-eekairi, thireekairi, rihi hiru*, and *hiru*, respectively.

Rodhigandu is a distinctive band placed at the center of the *kasabu boavalhu* with its multiple patterns and vibrant colors. The bands above and below the *rodhigandu* must be symmetrical to one another.

The image illustrates a conventional arrangement of lace bands on a *boavalhu* (Fig. 18).

1. *Mathee kairi (hiru)*: The first band at the neckline
2. *Bagiya*
3. *Ranhiru*: The second *hiru*
4. *Bagiya*: Same size as the first *bagiya*
5. *Ranhiru*: Same size as band 3
6. *Bagiya*: Same size as the previous *bagiya*
7. *Rihihiru*: Same size as the first band (*mathee kairi*)
8. *Rodhigandu*.

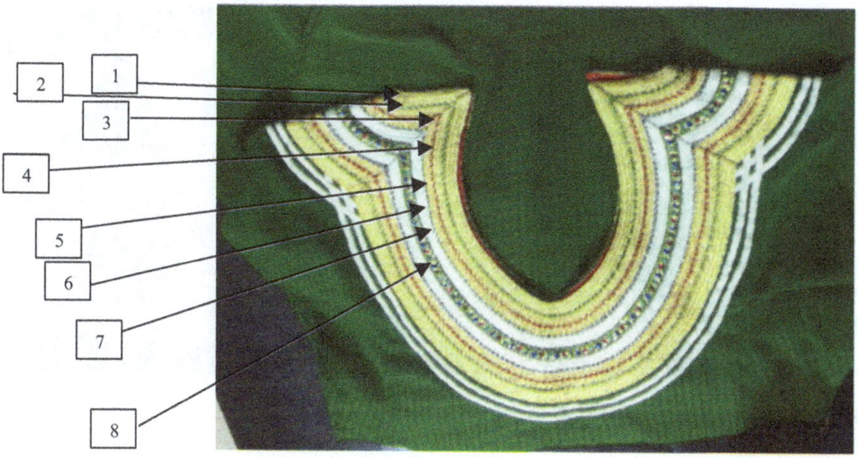

Fig. 18 Traditional kasabu libaas (lace fixed *libaas*). Photo: Aminath Abdulla, 2016

Once the *rodhigandu* (8) has been fixed to the dress, the sequence is repeated in reverse order. Thus, the final band (*hiru*) in the *boavalhu* will be 1 cm wider than the *mathee kairi* band (first band).

Materials required to weave the lace bands

1. Gold thread (embroidery thread, thick and thin)
2. *Firoashi* (tool used for thread reeling)
3. Silk thread
4. *Baadhalaa* (flat embroidery thread)
5. Pins
6. Cotton thread
7. Silver thread (embroidery thread, thick and thin)
8. Small knife/scissors (to cut thread).

The following tools are used to make the *kasabu boavalhu* (Fig. 19)

1. *Gathafai* (Round Pillow Stand)

   This stand is made of lacquered wood and is shaped like a round vase with a concave center. It is 7 in. high and 1 in. thick.

   Gathafai is hollow inside and has the same circumference at both extremes. The weaver can use the hollow inside to store finished lace bands and materials.
2. *Foalibai* (Bobbins)

   Foali is a 5-in. long, carved wooden bobbin with the *kasabu* (gold/silver thread) wrapped at one end of it. A weaver will use 34 of these *foalis* to make the lace bands for a *kasabu*, and they are collectively known as *foalibai*. The heavier the bobbin, the easier it is to weave, as the weight helps the threads hang better. Therefore, the *foali* tend to be made from heavy woods such as Ironwood. Finally, they are used to braid the *kasabu* into the desired patterns.

Fig. 19 Image of *Gathafai* (pillow stand), *Gathafaige boa* (pillow), and *Foalibai* (bobbins). Photo: Aishath Mohamed Rasheed, 2021

3. *Gathafaige boa*
   This is a spherical pillow placed on top of the *gathafai* stand. It has a circumference of 2.5 ft. and is covered with a white pillowcase. The *kasabu* is braided on this pillow.
4. *Baalees/Bondi*
   This is a small wooden rectangle (2 in. by 5/6 in.) with a thread wound around it. The tool is thinner at the center, enabling the thread to be wound more tightly. A special kind of thread known as *baadhelaai* (flat *kasabu*) is wrapped around this tool and held in place by a small wooden dowel threaded through eyelet hooks (Figs. 20 and 21).

Table 2 summarizes the characteristics of the different types of lace bands and the number of *foali* and length of thread required to weave them.

7 Conclusion

In conserving Maldivian heritage, which is under threat from rapid social, economic, cultural, and environmental change, cultural heritage is inextricably bound to natural heritage. Renewed interest in cultural products such as *thundu kunaa* shows that the revival of ancient traditions and knowledge can contribute to protecting biodiversity and heritage, generate value within the Maldivian economy, and provide a new source of income for women [32]. Many traditional practices are now at risk of disappearing, and, in some cases, the recovery of handicrafts and memory is primarily due to tourism. Tourism can play an essential role in this process. However, it is crucial for their survival that local handicrafts' high value and quality, which find themselves in competition with cheaper imported handicrafts, be effectively communicated via detailed information about all the stages in the production process. Forbes reported that when the first tourists arrived in the Maldives in the early 1970s, the number

Fig. 20 Bobbin laces woven on a *Gathafai*. Photo: Aishath Mohamed Rasheed, 2021

Fig. 21 Close up of *Kasabu* weaving. Photo: Aishath Mohamed Rasheed, 2021

Table 2 *Kasabu* weaving

Type of *Kasabu* (lace)	Materials needed	No of thread per bobbin	No of foali (bobbin)/design	Length of the thread (ft.)	Approximate length after weaving (ft.)
Mathee kiri (first lace)	Gold thread	5 thread	21 *foali* (bobbins)	5	3.9
Bagiya (flat colored lace)	Silk thread and gold thread	2 thread	16 *foali* (bobbins)	5	3.9
Ranhiru (gold color lace)	Gold thread	5 thread	17 *foali* (bobbins)	5	3.9
Rihi Hiru (silver color lace)	Silver thread	5 thread	17 *foali* (bobbins)	5	3.9
Rodhigandu (the center lace)	Silk thread and badhala thread	6 thread	20 *foali* (bobbins)	5	3.4
Thiree Kairi (lower lace)	Gold thread	5 thread	28 *foali* (bobbins)	5	3.2
Vah Bagiya	Silver thread	2 thread	20 *foali* (bobbins)	5	3.9

Source Shafeeg [31]

of weavers, already in decline, initially increased dramatically in the southern atolls due to the direct sale of mats [23, p. 21]. The subsequent segregation pattern and spatial concentration of the resorts around Malé and changes in local lifestyles led to the subsequent decline in weaving, culminating in its almost complete disappearance today. Currently, the newly established National Center for Cultural Heritage (2019), a governmental initiative mandated by the Ministry of Arts, Culture, and Heritage, is conducting a large-scale survey of the country's tangible and intangible heritage to "safeguard the perpetuation of objects and sites of historical significance for future generations and ensure the documentation, preservation, and protection of cultural heritage" [22]. Handicrafts are promoted via national exhibitions, such as Fannu Expo, organized by the Ministry of Economic Development to support the creative arts, or Dubai 2020 (postponed to March 2021) to present local products internationally. Another key initiative is the proposal of launching a national brand—Authentic Maldivian Product—to assist in the identification, promotion, and marketing of domestically produced handicrafts, along with the setting up of the Maldives Handicrafts Center: "The purchase of a "100% Maldives" branded product would appeal to tourists so that they know they are supporting the local economy" [14, p. 23].

In this context, the role of NGO.s and cooperatives is crucial in helping to connect local craft production with the resort market [33]. Small-scale, cooperative systems can help reduce overall transaction costs, provide assistance where resources are scarce, and act as intermediaries in the supply of raw materials and the marketing of finished goods. Similar to MACCS, NGO.s have also developed a range of new, user-friendly products that are easier to sell on the tourist market and deliver training to women wishing to get involved in the crafts sector. *Kasabu* weaving has also been revisited and adapted to contemporary fashion. The already mentioned local artist Hidha Latheef creates handmade accessories such as wrist bands and ornaments using traditional techniques. Other small businesses that promote local artisans through their heritage-related designs are "Island Bazaar" "Toddy" and "Oevaali art". The revival of traditional clothing has also been perceived as positively contributing to identity and is the object of lively discussion and exchange on internet forums (e.g., @TheLibaasMvmn, #LibaasMovement) (Figs. 22 and 23).

Change is always very rapid in the Maldives. New courses, private initiatives, and, hopefully, a degree course will soon significantly advance textile education in the Maldives and protect traditional textiles as heritage. The key question is

Fig. 22 Laptop bag and a table runner made from traditional matting. Photos: Dhimyan Ahmed, 2019

Fig. 23 *Kasabu* handicraft made by Hidhana Latheef. Photos: Hidhana Latheef, January 2019

whether the revitalization of ancient traditions and knowledge concerning ancient crafts and weaving and the preservation of island heritage can withstand the rapidity of the ongoing transformation process. As the examples outlined show, original craft production is time-consuming, labor-intensive, and demanding. Young people typically value traditional craftwork for its designs. However, very few of them learn traditional methods and continue to apply them as in the past. This is because the techniques involved are highly labor-intensive and need to be instilled into young minds at an early age. The Maldivian school system tends to keep children constantly engaged in curricular study with little time for extra activities such as handicrafts. Unlike the generation before them, craft workers who are parents often fail to involve their children in their work with a view to encouraging their interest. In light of lifestyle changes triggered by tourism, it remains to be seen whether the younger generations, having been exposed to different experiences and aspirations to their parents and grandparents, will be able to find sufficient cultural and economic motivation for continuing to engage in quality craftsmanship.

References

1. National Bureau of Statistics (2018) Statistical yearbook of Maldives 2018. Malé
2. MoT (Ministry of Tourism) (2020) Monthly statistics. Republic of Maldives, Ministry of Tourism
3. National Bureau of Statistics, Ministry of Tourism (MoT) the Republic of Maldives (2019) Employment in tourist resorts. Malé
4. Haun B (2008) Inventing 'Easter Island.' University of Toronto Press, Toronto
5. MOTAC (Ministry of Tourism, Arts & Culture) (2013) Fourth tourism master plan 2013–2017. Republic of Maldives, Malé
6. dell'Agnese E (2018) "One island, one resort". Il turismo enclave alle Maldive come eterotopia pianificata. Boll Soc Geogr Ital, 1(1):27–39. https://doi.org/10.13128/bsgi.v1i1.87
7. MoT (Ministry of Tourism) (2019) Annual publication. Ministry of Tourism, Republic of Maldives
8. Kothari U, Arnall A (2017) Contestation over an island imaginary landscape: the management and maintenance of touristic nature. Environ Plann A 49(5):980–998
9. M.E.E. Ministry of Environment and Energy (2015) Maldives. Climate change policy framework. http://www.environment.gov.mv. Accessed 1 Aug 2019
10. ICSU, International Council for Science (2002) Science. Traditional knowledge and sustainable development. In: Series on science for sustainable development, 4. ICSU. https://unesdoc.unesco.org/ark:/48223/pf0000150501. Accessed 27 Feb 2020
11. Mohammad S (2007) From the treasure of Maldivian traditional medicine. Guide 1. Word Offset Printer, Thrissur, Kerala
12. Jameel MM, Ahmad Y (2015) Architectural heritage of Maldives and its revival through tourism. In: Conference Islam and multiculturalism: Islam in global perspective, conference paper, Jan 2015. N.Y.U. Abu Dhabi
13. Pyrard de Laval F (1887 [1619]) The voyage of François Pyrard of Laval to the East Indies, the Maldives, the Moluccas and Brazil, translated into English from the third French edition of 1619, and edited with notes, by Gray A, Bell HCP, vol I. Council of the Hakluyt Society, London
14. Zakariyya A, Nisha M, Rasha A (2007) Handicraft in Maldives. Ministry of Economic Development and Trade, Malé

15. Young IA, Christopher W (1844) Transactions of the Bombay geographical society from 1836 to 1838. American Mission Press, Bombay, M. Gaekwad's Oriental series No 122, pp 54–86
16. Battuta I (1976) The Rehla of Ibn Battuta (India, Maldive Islands and Ceylon), translation and commentary by Husain M. Oriental Institute, Baroda
17. Maloney C (1976) The Maldives: new stresses in an old nation. Asian Surv 16(7):654–671
18. Bell HCP (1883) The Máldive Islands: an account of the physical features, climate, history, inhabitants, productions and trade. F. Luker, Acting Government Printer Ceylon, Colombo
19. Schmidt di Friedberg M (in print) Heritage e biodiversità alle Maldive tra scomparsa e commodification per il turismo sostenibile. I thundu kunaa (stuoie artigianali) dell'atollo di Huvadhoo. Geotema
20. Schmidt di Friedberg M, Abdulla A (2021) The gender dimension of environment in the Maldives. In: Malatesta S et al (eds) Atolls of the Maldives: nissology and geography. Rowman & Littlefield, pp 45–63
21. Rosset CW (1886) The Máldive Islands. The Graphic, 16 Oct 1886. Maldives Royal Family. http://www.maldivesroyalfamily.com/rosset.shtml. Accessed 7 Jan 2021
22. Abdulla A (2017) The production and design of thundu kunaa mats in Huvadhu atoll. Malé, MACSS
23. Forbes A, Ali F (1980) Weaving in the Maldive Islands, Indian Ocean: the fine mat industry of Suvadiva atoll (illustrated by mats held in the collection of the museum of mankind), occasional paper. British Museum
24. Mohamed N (2005) Note on the early history of the Maldives. Archipel 70:7–14
25. Knoll E-M (2018) The Maldives as an Indian Ocean crossroads. In: Oxford research encyclopedia of Asian history. https://doi.org/10.1093/acrefore/9780190277727.013.327. Accessed 15 Jan 2021
26. Romero-Frias X (2015) Between isolation and diffidence: possibilities and peculiarities in contemporary practices in the Maldives. In: Sociology and social anthropology in/for South Asia histories and practices. A publication oriented one-day Seminar, 12 Dec 2015. https://www.academia.edu/21281069/Between_Isolation_and_Diffidence_Possibilities_and_Peculiarities_in_Contemporary_Practices_in_the_Maldives. Accessed 22 Jan 2021
27. Rosset CW (1887) On the Maldive Islands, more especially treating of Male Atol. J Anthropol Inst G B Irel 16:164–174
28. Maloney C (1980) People of the Maldive Islands. Orient Blackswan, Hyderabad
29. SAARC, Cultural Center (2015) https://saarcculture.org/2016/01/27/handicrafts-4/. Accessed 2 Feb 2021
30. Khanna JM (2019) Tracing the rich history of Kerala's traditional white and gold kasavu saris. Vogue. https://www.vogue.in/fashion/content/the-history-behind-kerala-kasavu-saris
31. Shafeeg A (1988) Dhivehi Masaiykaiytherikan. In: Malas 15. National Center for Linguistic and Historical Research, Malé, pp 9–15
32. Kanvinde SH (1999) Maldivian gender roles in bio-resource management. F.A.O. Regional Office for Asia and the Pacific, Bangkok
33. Alm E, Johansson S (2011) Women in tourism; challenges of including women in the Maldivian resort sector. Lund University

Aminath Abdulla is the founder and chairperson of Maldives Authentic Crafts Cooperative Society (MACCS). The community-based organization in the Maldives focuses on reviving and promoting traditional handicrafts and the empowerment of women. Aminath is dedicated to strengthening the market for authentic local crafts and the livelihood of home-based workers through documentation, restorative and regenerative projects. She has published works on traditional mat weaving documenting the first written form of the craft, its origins, and its process. While contributing towards the intangible cultural heritage of crafts, she also serves on the advisory board of HomeNet South Asia and is a member of the Business Council, Ministry of Economic Development in the Maldives.

Marcella Schmidt di Friedberg is a full Professor of Geography at the "Riccardo Massa" Department of Human Sciences for Education, University of Milano-Bicocca (Italy). She is vice-director of the Marine Research and High Education Center (MaRHE) in Faaf-Magoodhoo (Rep of Maldives) and Chair from 2016 of the International Geographical Union (I.G.U.) Commission on History of Geography. Her research interests concern Cultural Geography, Hazard and Resilience, Gender Geography, and History of Geographical Thought. She has been working and publishing extensively on the relations between nature, culture, memory, and landscape in different contexts, from the Mediterranean to Japan.

Weaving Through Generations: A Study on the Transmission of Bhutanese Weaving Knowledge and Skills Over Three Generation

Joseph Lo and Pema Chhoden Wangchuk

Abstract This chapter accounts how three different generations of Bhutanese weavers, spanning some 60 years, learned their weaving skills. Their narratives are analyzed to understand how the transmission of weaving knowledge and skills has evolved (or not) and their thoughts in establishing the relevance and appropriateness of teaching this traditional practice to new generations of weavers to sustain this traditional practice.

Keywords Bhutanese textiles · Transmission of weaving knowledge and skills · Cultural sustainability · Practice-based training · Creativity and innovation

1 Introduction

1.1 Aims of This Research

The goal of the partnership between RTA and CFCH's project is to sustain the practice of weaving in Bhutan. Part of sustaining this practice is understanding how weaving skills and knowledge have been passed from one generation to the next. This understanding will assist in formulating credible programs that will perpetuate the practice of weaving. Therefore, this research aims to contribute to the sustainability of weaving practice in Bhutan.

J. Lo (✉)
Centre for Folklife and Cultural Heritage, Washington, DC, USA

P. C. Wangchuk
Royal Textile Academy of Bhutan, Thimphu, Bhutan
e-mail: pcwangchuk@rtabhutan.org

X. Yan et al. (eds.), *Textile and Fashion Education Internationalization*, Textile Science and Clothing Technology, https://doi.org/10.1007/978-981-16-8854-6_9

1.2 Objectives of This Chapter/Research

The objectives of this book chapter are to:

- Understand how weavers from different generations started their weaving journey and learned how to weave;
- Comprehend the contexts of their realities when these weavers started weaving;
- Appreciate the similarity or difference between the experiences of these weavers;
- Acknowledge their views on the transference of weaving skills to future generations of weavers and provide recommendations for the formulation of future weaving training programs for RTA.

1.3 Content of This Book Chapter

The focus of this book chapter will be the findings, analysis, and discussion section. This book chapter will briefly introduce Bhutan through its history and an overview of Bhutanese textiles. The following section will comprise a short description of the research methodology. This is to give the reader a sense of how the findings were reached. The profiles of those weavers interviewed will also be included, giving readers a background to appreciate their responses. The responses from the interviewees will be analyzed by examining their similarities and differences in responses and other related matters linked to the transmission of weaving skills. Finally, a section on recommendations will be included for consideration when establishing future weaving training programs.

2 History of Bhutan and Overview of Bhutanese Textiles and Weaving

2.1 Overview

The Kingdom of Bhutan, also known as 'Land of the Thunder Dragon', is a small Buddhist kingdom located in South Asia bounded by Tibet/China on its northern borders and to the south, east, and west by India (Fig. 1). The total land area is about 47,000 km^2, making it comparable in size to Switzerland [1, 9]. Presently, the total population of Bhutan is 748,931, and the average population density is 19.5 people per km^2 [6].

In Dzongkha (the official Bhutanese language), 'Bhutan' is known as 'Druk yul' (the Land of Thunder Dragon) and was founded by Zhabdrung Ngawang Namgyel (1594–1651) in 1616. However, it was not until the early 1900s that this region was again consolidated under Ugyen Wangchuck (1862–1926) [1]. He founded the Wangchuck Dynasty and was crowned the first Druk Gyalpo (Dragon King).

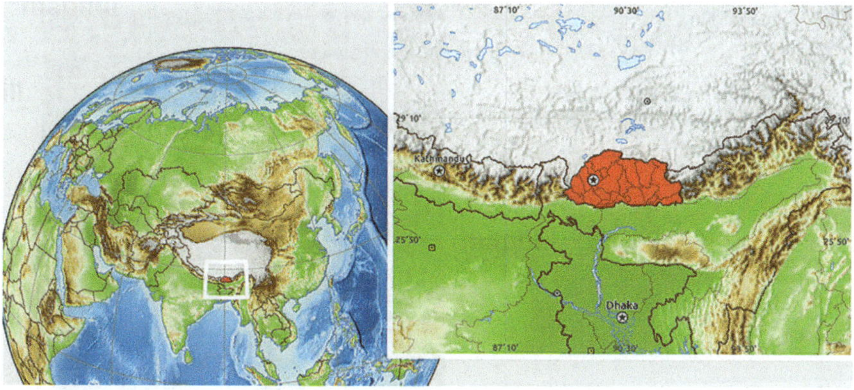

Fig. 1 Location and map of Bhutan and surrounding regions (Maphill.com)

From then on, Bhutan was ruled by a succession of Druk Gyalpos, who are his descendants. Today, Bhutan is a constitutional monarchy where the state's power is divided between the monarchy, government, and monastic bodies. The fifth Druk Gyalpo—Jigme Khesar Namgyel Wangchuck (b. 1980) is currently the Head of State.

Presently, Bhutan is the only Mahayana Buddhist Kingdom, with 70% of the population practicing Buddhism. Inculcating a distinct national philosophy based on Buddhism, Buddhist thoughts permeate all strands of secular life, influencing the country's cultural, ethical, and sociological development.

"Further, in a period when economic development and modernization were perceived as being synonymous with westernization, our country placed immense value on the preservation and promotion of traditional culture and values. For a small landlocked country, located between two large countries, our culture and traditions have provided the basis for our national identity and thus bolstered our security and sovereignty" [7].

Importantly, Bhutan's economy is also built upon Buddhist spiritual values, manifested in the concept of Gross National Happiness (GNH) [2].

2.2 *National Dress*

Infusing a national identity that defines Bhutan apart from its neighbors, Bhutan has a distinct national dress. 'Gho' is the national garment for men and 'kira' for women (Fig. 2). The gho is a male knee-length robe tied with a belt forming a pouch at the waist. Women wear the kira, a large rectangular cloth folded around the body, held together by brooches at the shoulders and belted at the waist, forming an ankle-length dress. A silk blouse—wonju—is worn under the kira while a short jacket—toego—is put over the kira [5]. Because of the fourth King's decree (kasho), wearing the national dress was made mandatory on 16 January 1989, national dress must be worn

Fig. 2 Gho and Kira

when visiting the dzongs (fortress housing government offices and monasteries) and for all public and formal occasions. Everyone from members of the Royal family to ordinary citizens must observe this decree except those operating modern machinery and those working outside the country [3].

2.3 History of Bhutanese Textiles

With Buddhism as its state religion, the few religious texts that survived many disastrous earthquakes and fires in the early seventeenth century documented Bhutan's history. Nothing, in general, was documented regarding textiles, social identities of weavers, or the country's history, but archaeological evidence suggests that the country was first inhabited during the Neolithic Period between 1500 and 2000 BC [5]. There are also records from early British explorers of their visits to Bhutan, folklore, and a few books that have survived. It was in the early seventeenth century when Zhabdrung Ngawang Namgyel introduced the gho that was later declared as the national dress for men in 1989.

3 Types of Bhutanese Textiles

During the first King Gongsar Ugyen Wangchuck's time weaving in Kurtoe, Lhuntse in the eastern part of Bhutan flourished. With noblewomen commissioning many elaborately designed kushutharas, their sense of style and fashion gained much prominence throughout the country. Within a short span of time, the attire gained much popularity and prestige.

The kushuthara (Fig. 3—Left) has evolved, leading to many variations in style and color. While motifs were made more complex and bolder, colored background kiras became a trend. It initially started with backgrounds of blue (ngosham) (Fig. 3—Right), red (mapsham), black (napsham) (Fig. 4—Left), and green (jangsham) to background colors such as pink, yellow, and orange. The newer design does not have a special local term and is categorized under pesar, meaning contemporary design (Fig. 4—Right). Like the kushutharas, pesar kiras display a stunning and striking impression when seen from a distance.

Aikapur, the supplementary warp patterned cloth that is traditionally woven out of bura (wild silk) or cultivated silk and occasionally of cotton, is another classic textile woven in eastern Bhutan, mainly in Radi, Tashigang. The cloth of choice for men as the aikapur is to a man's dress as a kushuthara is to a women's dress. Prized

Fig. 3 Kushuthara (left) and Ngosham (right) patterns

Fig. 4 Napsham (left) and pesar (right) patterns

Fig. 5 Aikapur (left) and lungserma (right) patterns

by Bhutanese for generations, this classic supplementary weft patterned weave has been the cloth of distinction saved for special occasions.

An aikapur (Fig. 5. Left) is featured with alternating bands of plain weave or 'pang' and supplementary-warp patterning or 'hor'. Traditionally all the variations of supplementary warp patterned band weaves were collectively referred to as aikapurs. However, today the aikapur is generally referred only to as a piece with a combination of white, yellow, and blue with jadrim, multi-colored stripes.

Mentsi mathra (Fig. 6—Left) is one example of this type of weaving with a red ground with golden yellow patterned bands. Other examples are lungserma (Fig. 5—Right) with green and red patterned bands on a yellow ground, aikapur with white, yellow pattern bands alternating with blue plain-weave bands, and jadrim with multi-colored stripes. Dromchuchema (Fig. 5—Right) is a combination of red, yellow, white, and green pattern bands. Moentha, which is only worn by women, has red and black pattern bands alternating with stripes of other colors, predominantly black and white.

Some versions of these fabrics are now made with pattern bands interspersed with single-faced supplementary weft patterning bands. One of the designs most admired is the tree motif or the shinglo. The skill of a weaver can be judged by how deftly the

Fig. 6 Mentsi mathra (left) and aikapur dromchuchem (right) patterns

weaver executes this pattern. The width and intricacy of the pattern bands on these textiles are linked with the prestige of the fabric and its intended wearer.

4 Current Context of Transmission of Weaving Skills in Bhutan

Presently, there are a few organizations (NGOs and private institutes) that conduct weaving training. They are listed below:

4.1 Royal Textile Academy of Bhutan (RTA)

The Weaving Department at the RTA provides training on weaving for a duration of two to three months. Three to four instructors are assigned depending on the number of participants. All skills related to weaving, from starching, dyeing of yarns, preparation of the warp, setting it on the loom, and finally weaving, are taught over two to three months. While there is no syllabus as such, the instructors follow a work plan which mentions the activities that will be carried out each week. As of June 2016, eleven batches of more than 250 trainees have completed the training.

4.2 Choki Traditional Art School (CTAS)

Established in 1999, CTAS is recognized and certified as a formal private training institute in Bhutan by the Ministry of Labour and Human Resources (MoLHR), Royal Government of Bhutan. CTAS provides training on five traditional craft courses, including weaving.

Ten slots are announced for the three-year weaving training, and usually, only three to four candidates come forward. Like other organizations and institutes, they do not have a syllabus yet, but the trainees learn all aspects of weaving under one instructor's supervision. Upon completing the training, the institute issues a diploma certificate or a National Certificate (NC) if the training fulfills the requirements and accreditation criteria set by the Department of Occupational Standards under MoLHR.

4.3 National Institute of Zorig Chusum (IZC)

The Weaving training at IZC is for two years. Ten to fifteen students attend the course with one instructor, who is a master weaver holding a bachelor's degree. Earlier, a diploma certificate was issued upon completion of the training, but now, the trainees are required to sit for a national-level assessment conducted by the Department of Occupational Standards under MoLHR to attain a National Certificate. IZC also does not have a syllabus yet, but they are working on one they hope to implement by the beginning of 2022.

4.4 SABAH Bhutan

SABAH, a non-profit organization, has trained about 750 participants in weaving so far but beginning 2019–2020, the organization could not conduct any training due to financial problems. However, this year—2021–10 women were trained in April. Their training usually lasts for a week because most of the trainees come equipped with basic weaving skills. Before 2019, the organization conducted training twice a month that was supervised by two instructors.

4.5 Tarayana Foundation Center, Agency of Promotion of Indigenous Crafts (APIC), and Handicraft Association of Bhutan (HAB)

Tarayana, APIC, and HAB are other non-profit organizations that have reached out to many women across the country and have transferred skills through workshops relating to new and old weaving techniques and entrepreneurship, design training, and dyeing the yarns the traditional way.

While there are no books, publications, or documentation on the weaving culture in Bhutan and, more importantly, on how the skill is transferred, it is almost entirely transferred as peer-learning or handed down within extended families. As a result, most weavers have learned their skills from friends or family and improved their skills through their individual efforts.

5 Research Methodology

The interviews were conducted under the auspices of RTA. Therefore, RTA's code of ethics was followed. The purpose and nature of this research were explained to weavers, and their consent was sought before the interview started. Their approvals were expressed through the signing of RTA's Interview Consent Forms.

The approach employed was to compare three weavers from different generations—youth, adult, and senior—and relate the methods they learned to weave. The data would then be compared to analyze the similarities or differences of how these three weavers from different generations learned their skills.

Following the age classification defined by National Statistics Bureau, this research will adhere to the same groupings. Thus, youth in the research will be defined as 15–24 years, adult as 25–64 years, and senior as 65 and above. The reason for this was to reflect and seamlessly link this data with the national framework, contributing to the intellectual resource of Bhutan.

The selection of the three interviewees was through a random selection methodology to identify and select weavers for this project. While it was easy to identify the youth and adult weavers in Thimphu (the capital of Bhutan where RTA is located and where the interviews would be conducted), the senior weaver was quite challenging as the majority of weavers in that age group (65 and above) were in the villages. The few ones who live in Thimphu were either not in town at the time of the interview or had stopped weaving from the time they moved to Thimphu from their villages.

A structured questionnaire with open-ended questions was constructed for this purpose. The reason for having a structured questionnaire was to facilitate cross-comparison between all three weavers. As such, the questionnaire was divided into the following sections:

- Personal details
- Family details
- Context when started weaving
- Genesis of weaving experience
- Current Practice
- Future.

As close-ended questions may prejudice interviewees' responses, this investigation decided to construct open-ended questions instead. This would enable interviewees to express themselves openly and freely without having to consider established responses. Assisted by the co-investigator (CI), the principal investigator (PI) drafted the questionnaire, subsequently reviewed by RTA and shared with CFCH.

In terms of the sequence of interviews, it was decided to interview the adult weaver first, and then talk with the youth, and finally, the senior weaver. This was because the adult weaver would set the norm from which deviations (if any) could be examined. The first interview was conducted on 23 March 2021, and the youth's interview was affected a week later, on 1 April 2021. Both these interviews were conducted at RTA's premises. The final interview with the senior weaver was conducted on 18 April 2021 in Eastern Bhutan when the CI was on a field trip there. This provided a rare opportunity to include a weaver that was not based in Thimphu.

As the PI could not be in Bhutan to conduct the interviews, the CI was on hand to conduct face-to-face interviews. However, using the digital communication service—Zoom—the PI participated in all the interviews as he directed the sessions while the

CI served as the coordinator and translator. Responses from the interviewees were recorded immediately by both the PI and CI. Upon completion of the interviews, the PI sent the completed questionnaires to the CI for verification. CI also photographed the textiles discussed by interviewees as examples of their work.

6 Profile of Weavers Interviewed

6.1 Interview 1—Youth

At the time of the interview, Ms. Dechen Tsering (not her real name) was 21 years of age. In 2018, she completed her education in Class 12 when she was 19 years old. Although her village is in Eastern Bhutan—Changshinggoenpa, Mongar, she currently resides in Lungtenphu, in Thimphu. Dechen Tsering just joined RTA's Weaving Training Programme and is currently learning the Trima technique.

In Changshinggoenpa (her village home), Dechen Tsering lived within a three-generation family structure, and amongst her siblings, she is the eldest. The family's main mode of production is farming, yielding an annual income of about BTN 36,000 or US $4816.08/- (exchange rate of 1 BTN = US $0.0134 on 26 April 2021). In addition, Dechen Tsering's mother weaves and sells one or two rachus (a traditional scarf for women) or kera (a traditional belt) per month. Thus, in terms of weaving's contribution to the family's income, it has not been significant.

6.2 Interview 2—Adult

Sonam Deki (not her real name) is an adult weaver, 44 years of age. She completed her formal education in 1992 at the age of 15, passing Class 8. Sonam Deki hails from Tashigang, Eastern Bhutan, from the village of Bartsham. Currently, she resides in Thimphu and has been here since 1999. Sonam Deki is currently employed as a permanent weaver at the RTA.

Sonam Deki identifies herself as a wife and mother in a household of four persons. Her husband works as a legal officer and is a salaried employee. The annual family income is estimated at BTN 850,000 or US $11,445.25 (exchange rate of 1 BTN = US $0.0135 on 27 April 2021), of which Sonam Deki contributes about 14% (approx. US $1259 or BTN 63,800).

6.3 Interview 3—Senior

Pema Choki (not her real name) is originally from and currently resides in Tsankhar, Radhi in Tashigang.[1] At the time of the interview, she was 74 years old. Pema Choki has been a widow for some years now, living alone. She has no formal education, relies on farming to meet her subsistence needs, and weaving has been her only source of cash income. According to Pema Choki, her annual income is about BTN 15,000 (approx. US $203/- Exchange rate of 1 BNT = US $0.0135 on 26 April 2021).

7 Findings, Analysis and Discussion

The responses from all three weavers showed that there are both similarities and differences. This book chapter examined these matters to draw out insights into their responses and include other thoughts raised during the interviews. Finally, the weavers also provided suggestions concerning how RTA could consider, incorporate, and improve their future weaving training program.

7.1 Similarities

Origins

By randomly selecting and identifying weavers for this research, one of the most obvious coincidences was that all the weavers interviewed were from Eastern Bhutan. For example, Dechen Tsering, the youth weaver, came from Changshinggoenpa, Mongar, the adult weaver's (Sonam Deki) ancestral home was in Tashigang while Pema Choki (senior weaver) grew up in Radi, which is also now her current home.

Weavers started their weaving journey in their teens

The other commonality between all the weavers interviewed was that their introduction to weaving was in their teens. Dechen Tsering and Pema Choki started weaving when they were 14 years old, while Sonam Deki was initiated when she was 15 years old.

[1] Radi is a small gewog (block) in Tashigang district best known for producing authentic hand woven bura (raw silk) textiles. Although the bura yarn is imported from Gudama, Assam, India, they dye it locally to produce the famous radipa buras, especially the aikapur (supplementary warp bands) patterned textiles.

These early teenage years were crucial because they were formative years to establish one's interest and develop one's appreciation for weaving.

Farming has been the families' mode of production

The family background of all three weavers was farming. All of them claimed that when they were growing up, their annual family incomes were low.

Agriculture contributes only 10% to GDP but accounts for 54% of employment. Moreover, working in agriculture is highly correlated with being poor: about 66% of poor household heads work in agriculture (worldbank.org).

This could possibly show the need for such families to seek additional income generation activities to enhance their income level. Hence, weaving might be a viable option.

The above analysis was confirmed when all three interviewees stated that additional income was crucial for them to take up weaving. Furthermore, from the longer-term perspective, all three weavers speculated that weaving would enable them to be financially independent—both Sonam Deki and Pema Choki's investment in learning to weave proved to have good returns. For example, Sonam Deki reported that she contributed to about 14% of the family's annual income stems (approx. US $1259 or BTN 63,800). As a result, she was able to purchase a fridge, a TV, and other items for the family. It has empowered her as she need not ask her husband for any money to buy things for the house. One of her proudest moments was when she was able to sponsor pujas (worship rituals) for three days through her earnings.

For Pema Choki, she has been a widow for some years now, living alone. She has no formal education but relied on farming to meet her subsistence needs. As a result, weaving has been her only source of cash income. According to Pema Choki, her annual income is about BTN 15,000 (approx. US $203/- Exchange rate of 1 BNT = US $0.0135 on 26 April 2021).

According to Dechen Tsering, the irony was that although these farming families were most in need of generating supplementary income through weaving, they did not have the time to weave. This was because farming took up most of their time and energy. Instead, wage-earning families already engaged with the cash economy were more involved in weaving as their women folks had more time to weave.

Another interesting prospect would be that as Bhutanese graduates from a less developed country to a middle-income country in 2 years (2023), more Bhutanese would be engaged with the cash economy. Would this imply that more people could be involved in weaving? Nevertheless, in many other studies, it has been the experiences of many communities that once they graduate from a lower economic status to a higher one, craft skills would be the first to be threatened. One possible reason for the endangerment of craft skills would be to participate in other income generation work that would be more efficient and less labor-intensive than making crafts.

Supportive and conducive environment for learning to weave

In all three cases, the immediate social environment of all three weavers was engaged in weaving; many of the interviewees' primary circles were weavers. For example, in the youth weaver's case—Dechen Tsering's mother weaved while her father and grandmother were pleased that she was interested in learning how to weave. Likewise, Sonam Deki's (adult weaver) two elder sisters and mother weaved while Pema Choki (senior weaver) was surrounded by weavers, including her paternal grandmother and neighbors. Hence, this daily exposure to weaving may have seeded their interest in weaving.

Furthermore, all three interviewees witnessed the economic potential of weaving. In Dechen Tsering's case, her mother sold her woven keras and rachus, from which she was able to acquire pocket money for herself. For the adult weaver—Sonam Deki—because of her family's economic circumstances, selling textiles that they made helped to supplement her family's income. Pema Choki observed the trading and bartering of bura textiles from an early age. According to her, when she was growing up, one sack of bura fibers (about 10 kg) was BTN 100. This cash would be obtained from selling previously woven bura textiles. Depending on the designs of the textiles, one bura textiles would sell for BTN 1000 then, making a significant profit.

Many of these people from the interviewees' inner circle also became their primary reference groups, mentors, and trainers; they nurtured the interviewees' weaving skills. For Dechen Tsering's case, she would sit with her mother and watch her weave as she was interested in weaving. Hence, because of her interest, her mother taught her how to weave.

Sonam Deki recollected that when she showed a keen interest in weaving, her mother assigned her tasks related to weaving, such as preparing the yarns, including winding, starching, etc. Then, realizing Sonam's interest in weaving, her sisters taught her the mathematical processes of warping—how to count the warp yarns and how to warp the loom. Besides warping, her sisters also taught her how to weave motifs and create patterns by counting and picking the warp yarns. The textiles (Fig. 7) are examples of Sonam Deki's works.

For Pema Choki's case, she recalled her grandmother teaching her how to weave Hor motifs and patterns (patterning through supplementary warp yarns). Her grandmother encouraged and supported her when she found it very difficult and wanted to give up.

Driving force behind learning how to weave

In all three cases, it was their families' economic situation that necessitated them to take up weaving to supplement their family income or become financially independent. However, participating in income-generating activities could also be achieved in a multitude of ways, for example, seeking employment in nearby towns. Hence, why did these three weavers choose to weave when there were other options? The book chapter proposes this was because of their passion and interest in textiles. Perhaps it was the beauty of the woven cloth that enthused them to take up and excel in this venture. This was evident during the interviews when they spoke about the textiles they saw during their formative years.

Fig. 7 Textiles woven by adult weaver (Sonam Deki). The one on the right is her recent kushuthara pesar

Learning while doing

Both Dechen Tsering and Sonam Deki stated that they just 'got into the loom and started weaving'. Pema Choki was initiated through spinning, yarn preparation, and then warping, but she also did get 'into the loom' and started weaving. As reported by Pema Choki, this methodology resulted in much wastage of yarns. Nevertheless, significantly, this practice-based learning provided a sound grounding for developing their weaving skills.

This practiced-based training was important, as expressed by Dechen Tsering. She shared that it was essential to be constantly weaving or else one would forget easily.

Pride and self-satisfaction upon completion of the first piece of textile

Another common finding amongst all three weavers was that they expressed pride and self-satisfaction upon completing their first piece of textile. Dechen Tsering recalled that the first piece of completed textile she weaved was a tablecloth (size: Width—30 cm, Length—119 cm). It took her about one week to complete. This was used during her family's annual puja (ceremonial worship). She remembered feeling a sense of pride upon completing this piece of work; this piece of textile is still in her possession back in her village. Examples of Dechen Tsering's works are displayed in Fig. 8.

Pema Choki's first completed piece was a plain weave pangtsi (plaid) bura—Karchang Gho—for sale. At the start of weaving each textile panel, Pema Choki would feel anxious, wondering how and when she would be able to complete it. However, upon completion, she felt happy and a sense of self-satisfaction. In total,

Fig. 8 First textile (left) and recent textile (right) woven by youth weaver (Dechen Tsering)

Pema Choki thought that it took her about a month to complete the pangtsi. Then (more than 50 years ago), she sold it for less than BTN 1000. Today, a similar piece of textile will cost BTN 17,000, which is approximately USD 238 (Exchange rate of 1 BNT = USD 0.014 on 13 May 2021). In Fig. 9, Pema Choki proudly shows off her recent textiles.

Fig. 9 Senior weaver (Pema Choki) with the textiles she recently wove

Current practice—still weaving

All three weavers reported that they are still currently practicing weaving. Dechen Tsering practiced weaving throughout last year. During this period, she wove eight pieces of plain weave kiras. Two of these were commissioned while she kept the rest. Out of these, she gave two to her sister.

In the past 12 months, Sonam Deki wove four different types of Bhutanese textile. These were a gho textile in teri-cotton and three panels of religious textiles (silk motifs/patterns on cotton-based). The gho textile was for her husband, while the religious textiles were part of her work at RTA.

Despite Pema Choki's age, she reported that she still practices weaving to sell her textiles. Thus, weaving has been her primary means of income generation. Weaving, in some ways, saved her life as she has been alone since her husband's passing. In the past 12 months, Pema Choki wove every month. During the day, she would spend four to five hours weaving.

Skills levels—above average to excellent

Another common trait amongst all the weavers interviewed was in their own assessment of their weaving skills quality. When asked how they compare their skills with others, they evaluated their skills ranging from above average to excellent.

The youth weaver—Dechen Tsering—considered her weaving skills above average, while the Sonam Deki, the adult weaver, said she was excellent. On the other hand, Pema Choki assessed her skills to be between above average but below excellence.

This is significant because it could be related to the constant practice of their weaving skills, as cited above.

Opportunity cost of weaving—none

In the interview questionnaire, one of the questions asked was: Do you think that weaving has stopped you from developing yourself further and/or in other areas? All the weavers replied that weaving did not stop them from developing themselves further, but instead, it helped them fulfill their most significant potential.

In Dechen Tsering's case, it was weaving that brought her out of her village to Thimphu to enroll with the RTA's weaving program. Thus, weaving provided her with the opportunity to see the world beyond Changshinggoenpa, Mongar. In Sonam Deki's opinion, weaving has not stopped her from developing herself further and/or in other areas. As a matter of fact, weaving has helped her realize her highest potential as she always felt that weaving was in her veins. Finally, for Pema Choki, weaving saved her life. After her husband's demise, she was left all alone, but thankfully, she has been able to sustain herself with her weaving skills without having to depend on anyone else.

7.2 Differences

The difference in their responses could be attributed to the regional variations in weaving techniques, practices, and culture exercised in Bhutan, analyzing the responses from the three weavers. For example, because Pema Choki was from Radhi, the main weaving material was bura—raw silk. However, Sonam Dekki and Dechen Tsering shared a more similar weaving tradition. Therefore, their weaving practices had more similarities with each other than with Pema Choki. These differences would be examined under the following subjects:

Yarns

Because of the presence of bura fibers and the tradition of weaving bura textiles in Radhi, Pema Choki learned to weave by using bura yarn. But as bura was not the tradition elsewhere, both Dechen Tsering and Sonam Deki used poly-cotton (locally known in Bhutan as teri-cotton) when learning to weave.

Initiation

As a result of these traditional differences, especially in the employment of fibers and yarns, the weavers' initiation into weaving have been different. For example, because poly-cotton came pre-dyed and in cones, Dechen Tsering and Sonam Deki only had to prepare the yarns by winding them into balls for weaving. Instead, both shared that they started learning how to weave by warping up the loom.

However, in Pema Choki's case, her foray into weaving was somewhat different. Bura fibers purchased from India would need to be processed into yarns in the village. Thus, Pema Choki first learned how to prepare the yarns by first boiling the fibers in ash water, spinning the fibers into yarns, and winding up the yarns into balls to prepare for weaving. In addition, she also learned how to dye the yarns into various colors using lac, turmeric, madder, indigo, walnut shells. It is only after learning these tasks that she picked up warping the loom.

Challenges

In terms of challenges, the three weavers shared that they faced different sets of difficulties. For example, when Dechen Tsering started weaving, the one task she found most difficult was picking—to create a clear shed to push through the shuttle. This could be attributed to Dechen Tsering being the youngest weaver and was still learning how to weave more complex designs.

As for Sonam Deki, she declared the tensions of the warp and weft must be very tight. Weft yarns must be beaten down hard to make a dense textile. This would make the textile stiff, a texture that is preferred to show the folds and maintain the shape of the kira when worn. Lastly, a densely woven textile also lasts longer.

Nevertheless, according to Sonam Deki, the challenge was to balance the tensions placed on the warp and weft yarns with the density needed to weave the motifs and patterns. If the tension were too tight, it would distort the shapes of the motifs and patterns. To achieve this balance between the high density of the woven textile while maintaining the shapes of the motifs and pattern would be tricky.

Pema Choki recalled that when it came to weaving Hor motifs and patterns (patterning through supplementary warp yarns), she found it very difficult and wanted to give up. Thankfully, she persisted, and today, this is her most accomplished weaving form. She has been able to weave the most complex of designs using hor (crosshatch technique).[2]

First completed piece

Because of the differences in weaving tradition and practices, the first pieces of textile woven by the three interviewees were diverse. For example, Dechen Tsering's first completed piece of work was a tablecloth (size: Width—30 cm, Length—119 cm). It took her about one week to complete. This was used during her family's annual puja (ceremonial worship).

Sonam Deki's first completed piece of textiles was a kera (narrow belt) which she wove when she was still in school. She could spend only 5–10 min every day because she also had to do her homework and household chores.

Pema Choki's first completed piece was a plain weave pangtsi (plaid) bura—Karchang Gho—for sale. In total, Pema Choki thought that it took her about a month to complete the pangtsi. Then (more than 50 years ago), she sold it for less than BTN 1000. Today, a similar piece of textile will cost BTN 17,000, which is approximately USD 226.36 (1 USD = BTN 75.1 as per Bhutan National Bank, 5/6/2021).

Training

In terms of being interested in participating in future training, both the youth and adult weaver said they were keen to partake in such activities if these opportunities arose in the future. Both would like to improve in the following areas:

- Advanced weaving
- Color theory
- Textile design
- Materials and yarn knowledge
- Natural dyeing.

Significantly, both were not opposed to paying for these training sessions. However, because of Dechen Tsering's limited financial resources, she would only be able to pay a nominal fee. For Sonam Deki's case, she viewed such training opportunities as a means of self-investment, and also, because of her financial situation, she said she was willing to pay for these training courses.

Because of Pema Choki's age, she said she was not interested in participating in any training to improve her weaving skills.

---

[2] *'Hor'* or warp pattern bands are characterized by their supplementary warp floating technique that forms bands of repeated motifs on a warp faced plain-weave ground. The number *'Kam'* or cross hatches in each supplementary warp pattern band is an indicator of the superiority of the textile. The number the cross hatches is always an odd number; usually seven but more elaborate weaves may have nine, eleven and even up to fifteen.

Others

Currently, Sonam Deki, the adult weaver, has been teaching her daughter to weave; her daughter just completed class 12 (19 years of age). Sonam Deki reasoned that weaving would be a useful skill to acquire because she would be able to weave and produce her own kira; furthermore, it is also a productive means of using her free time. Moreover, it would enable her to generate some 'pocket money' too.

Pema Choki reflected on how weaving today has evolved into a much simpler form. Today, yarns come ready-spun. Hence, it would not be necessary to spin bura fibers into yarns now. In Pema Choki's view, the quality of machine ready-spun yarns and handspun yarns were the same.

Pema Choki felt that some 50 years ago, circumstances made weaving challenging. For example, there was no electricity, and she weaved by candlelight. Today, all houses in Radhi have electric lighting. Hence, more time can be devoted to weaving. As a result, the quality of woven textiles should be better than those in the past. Thus, young weavers should take advantage of this to improve their work and be more creative and innovative.

8 Recommendations for Future Weaving Training Programs

In Dechen Tsering's view, she felt that how weaving has been taught needs to be re-considered to attract more young people; presently, it has not been interesting and would not sustain the interests of young people. Furthermore, Dechen Tsering stated that teaching weaving in stages might be more beneficial than asking one to sit in the loom and weave immediately—as her mother taught her. Another suggestion from Dechen Tsering was that filming each stage of the weaving and uploading it onto social media platforms would also be helpful. These could serve as records and references, enabling learners to view when they were in doubt.

Significantly, to complement such e-resources, Dechen Tsering shared that practice-based learning would still be very important. Having a weaving instructor by one's side would be effective, as problems faced by the trainee could be addressed immediately. Thus, the ratio between instructors and students would be an important consideration. Moreover, weaving instructors would need to be professional.

Dechen Tsering also added that innovation and creativity in weaving would be significant. She reasoned those new designs could be created rather than offering the same old designs to the market. It would be essential to give consumers more options through diverse choices and designs of textiles.

Sonam Deki stated that the most significant lesson in weaving would be color theory and how to combine colors harmoniously. In her opinion, even if the motifs and patterns were not well designed, if the color combination were good, the textile would sell. Therefore, the matching of color could be a very important lesson for textile designers.

Sonam Deki reminisced that during those formative years when she was learning to weave, her inspiration for textile colors came from nature. She would use color pencils to help design the colors of the kiras to mirror the colors of the rainbows. Thus, if future lessons could teach trainees how to translate and derive colors from nature into textile designs, it would be incredibly beneficial to the future of textile designing in Bhutan.

The basis for all future weaving training would still be practice-based training. Sonam Deki cited that the only way to learn about the balance of tensions between the warp and weft yarns while weaving motifs and patterns could only be experienced by sitting at a loom and weaving. It would be almost impossible to learn about the balance of tension without feeling it.

Sonam Deki revealed that she did not learn about the different types of fibers and yarns and their properties when she started weaving. No one was available to teach her natural dyeing techniques or exposed her to different types of textiles beyond traditional Bhutanese textiles. In terms of existing traditional weaving techniques, tools, and equipment, she only knew of the existence of horizontal frame looms through pictures of them in her school textbook; she knew that these types of looms were used in Central Bhutan. However, no one in her village used them or, for that matter, knew how to use them. Thus, it would be good if these could be incorporated into future weaving training programs to broaden the horizons of future weavers. Sonam Deki said she would like to learn how to weave on a horizontal frame loom regarding her own self-development.

When asked if Sonam Deki would like today's youth to learn the same way she did, she responded with a resounding 'no'. She said that she did not even have her own loom when she started but had to use her sisters' instead. Furthermore, today, there are so many ways of learning how to weave, which might be more efficient and effective.

Importantly, as she has been a weaving instructor at RTA, her opinion was that creativity and innovation need to be included and encouraged as part of the learning process. She believed that it would be very important to make herself available to her students so that technical problems need to be addressed immediately.

When asked to reflect on her learning experience, Pema Choki felt that her grandmother's attitude towards encouraging her to weave was very good. For example, in many instances, it was common to hit trainees when they made mistakes physically. However, in Pema Choki's case, her grandmother was very supportive and encouraged her to stay on course, even though it was difficult. Thus, she felt that being positive and supportive were good ways of training new weavers.

9 Conclusion

As implied in the title of this book chapter, one of the tacit assumptions of this research was that through examining the weavers from different generations, one would realize that the transmission of weaving skills has evolved. This evolution process would also be compounded through the changing context of society. Hence, three weavers from different generations were selected for this study. Dechen Tsering, the youth weaver who started weaving in the 2010s, while the adult weaver—Sonam Deki—learned to weave in the 1990s. Pema Choki, the senior weaver, honed her skills during the 1960s. Thus, theoretically, this study spans some 60 years.

However, upon analyzing the data from the interviews, there were more similarities than differences amongst the learning techniques, contexts, and experiences of the weavers, despite the differences of eras. For example, all three weavers started on their weaving journey during their teenage years. They all came from a rural background, and their families were not economically well-off. Yes, despite this, they grew up in an environment that was rich in hand-woven textile practice, and their interest in learning weaving was well supported and encouraged by their families. Significantly, their trainers and mentors during those formative years were members of their immediate family. Coincidentally, they all came from Eastern Bhutan, where the weaving culture has been well-grounded.

Nevertheless, aside from their marginalized economic situation, their interest in textiles and weaving propelled them to take up this craft, making it their primary income generation activity, even to this day. Interestingly, practiced-based learning was their primary method of learning. In terms of pride and self-fulfillment, all three weavers reported that it was most acute when they completed their first piece of work. Because of their constant practice, their self-assessed skill quality was high—above average to excellent.

Significantly, the difference in their learning experiences could be attributed to the regional differences in weaving techniques and practice. For example, Dechen Tsering and Sonam Deki shared the same weaving tradition, while Pema Choki's weaving tradition was focused on bura yarns and textiles. Thus, she learned how to prepare and spin bura fibers into yarns and learned different natural dyeing methods before learning how to wrap up the loom.

In terms of their views on the future of weaving training, Dechen Tsering and Sonam Deki shared some pertinent views. Because of Dechen Tsering's youthful age, she advocated for the use of multimedia to serve as a resource for learning. This will help to align weaving with the disposition of youths. Sonam Deki felt that color theory and the methodologies of translation of colors onto textile designs would be a very important module of the training, considering that it is the colors of textiles that would attract customers. Sonam Deki also advocated that the training syllabus should stretch beyond traditional Bhutanese textiles. For example, it should also include learning about different fibers, yarns, looms, and weave structures to broaden the horizons of Bhutanese weavers. Both Dechen Tsering and Sonam Deki agreed that it would be vital to include innovation and creativity in the learning of weaving to offer a diversity of textiles to customers. Furthermore, they both concurred that practice-based learning would be the best methodology to inculcate weaving skills. Significantly, from Pema Choki's perspective, it would be important to take a 'soft' approach to teach new weavers learning skills.

The aim of RTA and CFCH's project is to sustain the practice of weaving in Bhutan. One means of sustaining this practice is through training of new weavers to partake in the weaving traditions of Bhutan. Therefore, the formulation of weaving training programs to incorporate the recommendations from practitioners is important, especially in aligning traditional practice with the dispositions of today's young people. Therefore, this study has shown how weaving skills and knowledge have been transmitted over the generations and how these can be adapted and evolved to make learning relevant, credible, interesting, and useful. In doing so, it is anticipated that Bhutanese weaving traditions will evolve but also perpetuate through generations to come.

Acknowledgements This study was founded upon the collaboration between the Royal Textile Academy of Bhutan (RTA) and the Smithsonian Center for Folklife and Cultural Heritage (CFCH). The research team would like to thank everyone involved in assisting in this inquiry, especially those weavers who were interviewed; they gave their time generously in sharing their stories and experiences on the genesis of their weaving journey while recommending ways in which the transmission of weaving skills can be improved.

References

1. Aris M (1994) The raven crown. Serindia Publication, London
2. Centre for Bhutan Studies (n.d.) Gross national happiness. Available at: http://www.grossnationalhappiness.com. Accessed 23 May 2013
3. Hutt M (2005) Unbecoming citizens: Culture, nationhood, and the fight of refugees from Bhutan. Oxford University Press, New Delhi
4. maphill.com (n.d.) Bhutan: location. Available at: http://maps.maphill.com/bhutan/location-maps/physical-location-map-of-bhutan.jpg. Accessed 18 July 2014
5. Myers K, Bean S (1994) From the land of the thunder dragon: Textile arts of Bhutan. Serindia Publication, London
6. National Statistics Bureau (n.d.) Bhutan at a glance. Available at: https://www.nsb.gov.bt/publications/insights/bhutan-at-a-glance/. Accessed 5 May 2020
7. Royal University of Bhutan (n.d.) His majesty the King's address at the 14th RUB convocation, 24 May 2019. Available at: https://www.rub.edu.bt/index.php/en/the-university/chancellor1/588-his-majesty-the-king-s-address-at-the-14th-rub-convocation-24-may-2019. Accessed 5 May 2021
8. The World Bank (n.d.) The World Bank in Bhutan. Available at: https://www.worldbank.org/en/country/bhutan/overview. Accessed 13 May 2021
9. Wangdi K (n.d.) Bhutan: country pasture/forage resource profiles. Available at: http://www.fao.org/ag/AGP/AGPC/doc/Counprof/Bhutan.htm. Accessed 21 May 2013

Joseph Lo A curator with Centre for Folklife and Cultural Heritage, Smithsonian Institution, Washington DC, Dr. Joseph Lo is working on a project with the Royal Textile Academy of Bhutan, Thimphu, to sustain and re-capitalize Bhutanese weaving practices. Dr. Lo was graduated from the School of Textiles and Design, Heriot-Watt University, Scotland, in 2015; Dr. Lo's academic contribution has been to determine self-identified markers of authenticity in the work of artisans. Practically, it addresses the balance between protection and conservation of the community's identity with openness to design innovation and development from outside.

Dr. Lo has worked for numerous UN agencies on culture and development, specifically focusing on culture-based creative industry, artisan development, and cultural diplomacy. Dr. Lo also sat on the Board of Advisors for British Council's Crafting Futures program for East and Southeast Asia (2017–2019) and is currently acting as an advisor to World Crafts Council (WCC) International and the Asia–Pacific Region.

Responding to the Covid-19 pandemic, Dr. Lo drafted a framework for WCC to build greater resilience in the crafts sector—a means to mitigate future crises. This resulted in Dr. Lo curating, coordinating, and organizing a global webinar to discuss this matter.

Dr. Lo has over two decades of first-hand experience in Afghanistan, Bangladesh, Bhutan, Cambodia, China, Indonesia, Laos, Mongolia, Myanmar, Nepal, Pakistan, Thailand, Tibet Autonomous Region, and Vietnam.

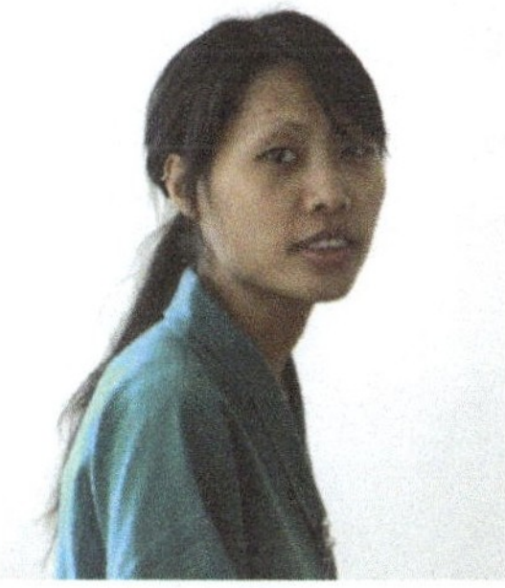

Pema Chhoden Wangchuk serves as a Senior Curator at the Royal Textile Academy of Bhutan (RTA), the first of its kind in the country that focuses primarily on preserving and promoting the traditional art of weaving. Established in June 2013, the museum has about 2000 pieces in its collection, mostly hand-woven textiles. Pema served as the museum curator for the last ten years since 2010, and just a few months ago, she moved to the Weaving Department. Pema also worked as a teacher for three years in a Montessori school dealing with children of 2–5 years before working with the academy.

With the museum concept in Bhutan still at an infant stage, Pema is one of the country's only three trained museum professionals. Specializing in Designing and Curating exhibitions, she has been involved in several exhibitions and events held within and outside of Bhutan. She led the exhibition team to Ahmedabad, India, where Her Majesty, the Queen Mother of Bhutan, graced the convocation ceremony of the National Institute of Design (NID), India. She also served as the chief coordinator for the exhibition 'Bhutan; Land Close to Heaven' that was on view for nine months at the National Museum in Prague, Czech Republic. The last international exhibition that she worked on was during the 'Bhutan Week' festival held in IGNCA in New Delhi, 2017.

She had internships in a number of museums in the US and Europe, trying to understand how efficiently each museum functioned. Pema played a significant role in the overall set-up of the Royal Textile Academy. She prepared and presented the 'Transition and Moving Plan' to the RTA Board that ensured a swift transition of the old Textile Museum into the bigger RTA premises. While equipping the collection's storage unit with a new CSS (Collections Storage System), Pema also coordinated the first exhibition Catalog book titled 'Thagzo; The Textile Weaves of Bhutan' book during the inaugural of the academy.

Designing and conducting Education programs for the RTA was another project which Pema spearheaded. As a newly established museum in the country, it was equally important to involve the younger generation to understand the primary purpose of museums. Pema also serves as a member of the Weaving Committee at the academy. She oversees the management, provides guidance, and takes an active part in decision-making while ensuring the smooth functioning of the program.

She has obtained certificates in 'Strategic Leadership for Schools in Changing Environment' from Sona School of Management in Salem, South India, and "Train-The-Trainer In Home-Based Care For Young Children" from the Regional Training and Resource Center in the Early Childhood Care and Education for Asia in Singapore. Ms. Pema holds a BA in Economics from the University of Pune, India, and an MA in Museology from the University of Washington, Seattle.

Correction to: Textile and Fashion Education Internationalization

Xinfeng Yan, Lihong Chen, and Hafeezullah Memon

Correction to:
X. Yan et al. (eds.), *Textile and Fashion Education Internationalization*, Textile Science and Clothing Technology, https://doi.org/10.1007/978-981-16-8854-6

In the original version of the book, the following belated corrections were incorporated:

The book's subtitle were updated from "A Promising Discipline from Southeast Asia" to "A Promising Discipline from South Asia".

In Chapter 4, the author name was changed from "Muzzamil Hussain" to "Muzammal Hussain".

In Chapter 7, all authors' affiliations were corrected.

The book and the chapters were updated with the changes.

The updated version of this book can be found at
https://doi.org/10.1007/978-981-16-8854-6
https://doi.org/10.1007/978-981-16-8854-6_4
https://doi.org/10.1007/978-981-16-8854-6_7

X. Yan et al. (eds.), *Textile and Fashion Education Internationalization*, Textile Science and Clothing Technology, https://doi.org/10.1007/978-981-16-8854-6_10

The manufacturer's authorised representative in the EU is Springer Nature Customer Service Centre GmbH, Europaplatz 3, 69115 Heidelberg, Germany. If you have any concerns regarding our products, please contact ProductSafety@springernature.com

Printed and bound by CPI Group (UK) Ltd, Croydon, CR0 4YY

30/09/2026

02225332-0002